Relativity for the Questioning Mind

Relativity

for the Questioning Mind

Daniel F. Styer

The Johns Hopkins University Press
Baltimore

The Johns Hopkins University Press
2715 North Charles Street
Baltimore, Maryland 21218-4363
www.press.jhu.edu

ISBN 13: 978-0-8018-9759-7   (hc)
ISBN 10: 0-8018-9759-9   (hc)
ISBN 13: 978-0-8018-9760-3   (pbk)
ISBN 10: 0-8018-9760-2   (pbk)

Library of Congress Control Number: 2010926450

*Special discounts are available for bulk purchases of this book. For more
information, please contact Special Sales at 410-516-6936 or
specialsales@press.jhu.edu.*

The Johns Hopkins University Press uses environmentally friendly book
materials, including recycled text paper that is composed of at least 30 percent
post-consumer waste, whenever possible. All of our book papers are acid-free,
and our jackets and covers are printed on paper with recycled content.

To all those
friends, relatives, and colleagues,
"familiar with life,"
who helped our family
during my wife Katie's illness and death,
and especially to
Kathy Abromeit, Ed Miller, and Joanie Webster

We should take comfort in two conjoined features of nature: first, that our world is incredibly strange and therefore supremely fascinating . . . second, that however bizarre and arcane our world might be, nature remains comprehensible to the human mind.

—STEPHEN JAY GOULD, *DINOSAUR IN A HAYSTACK*

Contents

Invitation

Two themes run through this book. First, that our universe is spectacularly bizarre, and deliciously avoids being bottled up by our commonsense notions of how it *ought* to behave. Second, that even in this, our non-commonsensical universe, observation and reason combine to form a powerful tool—the most effective means ever devised for probing the unknown. The topic that carries these two themes forward is Albert Einstein's theory of relativity, one of the finest products of human intelligence and imagination.

The importance of relativity has not gone unnoticed: On 31 December 1999, *Time* magazine declared Albert Einstein to be the "person of the century." The U.S. Library of Congress owns 897 books about relativity, many of them excellent, and 19 by Einstein himself. How can this slim volume hope to add anything to such illustrious company?

Nearly all of those 897 books fall into either of two categories: descriptive or technical. The descriptive books talk about the strange and beautiful results of relativity: They describe how moving clocks tick slowly, how time is connected to space, how moving rods shorten, how moving balls become more massive. This approach can give the misimpression that relativity is a set of unrelated and incomprehensible facts, like a collection of parlor tricks that work through sleight of hand, rather than a coherent picture of how our universe works. Even more significantly, descriptive books tell you what these phenomena are without telling you how we know about them. They expect you to accept such counterintuitive facts on the basis of the author's authority. I don't want you to accept facts on the basis of authority—whether that authority is a professor, a president, or a Führer. I want you to actively participate in building your own knowledge of relativity. As the title of this book emphasizes, I want you to question everything I tell you.

Question: If I question you, how will you answer me?
Answer: Like this.
Q: But how will you know my questions?

A: I anticipate your questions by critically evaluating my own understanding of relativity and by looking at treatments in other books, but most of all by listening carefully to the questions that my students have asked me. Indeed, in recent years I have required each student in my relativity classes to write out and submit to me a question at the end of each class meeting. This has given me a lot of insights and a lot of great questions.

Q: By encouraging me to question, aren't you undermining the authority of Einstein?

A: On the contrary, I'm encouraging Einstein's attitude that "a foolish faith in authority is the worst enemy of truth."*

Please don't think that I am disparaging descriptive books—when I was a teenager, my reading of Lincoln Barnett's descriptive *The Universe and Dr. Einstein* helped change the shape of my life—but *Relativity for the Questioning Mind* is analytic rather than descriptive.

I am also indebted to many technical books on relativity, but this is not a technical book. For one thing, the technical books often assume a rich understanding of mathematics, and I want this book to open up the relativistic world to those who know only simple algebra and the Pythagorean theorem of geometry. But there is a more important reason. Mathematics is a tool of extraordinary power, yet this very power can be used as a curtain to avoid any sort of understanding while simply solving problems. I know—I abused mathematics in this way for two decades. An analogy is helpful here: A gun is a more powerful hunting tool than a bow and arrow. If your hunting objective is simply to bring home game, then you use a gun. But if your hunting objective is to learn about animals, to understand their habits so that you can approach them closely, to grow familiar with the forest and its inhabitants, then the bow and arrow is the superior tool. The very power of the gun means that you don't need to understand the animals. For exactly this reason, this book deliberately avoids the use of powerful mathematics as a tool for solving problems without thinking or understanding.

I have no desire to disparage technical books, and even less desire to disparage the deep and subtle beauties of mathematics. Covariant and contravariant components, the stress-energy tensor, the Poincaré group—the names alone make me shiver in delight. Any professional must master such mathematical tools both in order to solve problems reliably and efficiently, and in order to extend the ideas to topics like mass

*Walter Isaacson, *Einstein: His Life and Universe* (New York: Simon and Schuster, 2007), p. 22.

and energy that we will not treat here. But such professional training is *not* the job of this book.

To summarize, *Relativity for the Questioning Mind* is rigorous but nontechnical. It is intended for a general audience, that is, for both scientists and nonscientists. It does not use advanced mathematics. It does not introduce the terminology or the algorithms that a professional must master. Given these limitations it cannot discuss all interesting topics, but it seeks to be analytic, authentic, and 100% accurate— rather than to "give the correct impression"—in those topics that it does treat. There are three occasions in this book (pages 20, 96, and 158) where I cannot achieve this lofty goal, and I need to tell you facts without thoroughly exploring the experiments and analyses that underlie the facts. I warn you in all three cases.

I thank, first, the 2,776 students who have, since 1989, taken my Oberlin College courses on relativity. Their struggles, enthusiasms, insights, criticisms, suggestions, and questions improved my own understanding enormously. One of these students, Victor Wong, particularly forced me to recast my arguments into a more tactile and robust form. Victor Wong is blind. I never would have finished this book were it not for a research status leave from Oberlin College. My friends, the physicist Edwin Taylor and the poet Marci Janas, read the manuscript carefully and made numerous helpful suggestions. In the last month of her life, my wife Katie encouraged me to complete this project.

PART I / Moving

CHAPTER ONE

The Paradox of the Mirror

I n 1894, when he was 16 years old, Albert Einstein wrote to his uncle Cäsar Koch, apologizing for being "a terribly lazy correspondent" but then describing what was going on in his life. He had been pondering questions like this one: "If I were in a railroad coach moving at the speed of light, and I looked into a mirror, then what would I see?" a quandary that we'll call "the paradox of the mirror." He closed by offering "warm greetings to dear aunt and your lovely children."

Let's think about this situation a little before trying to come up with a tentative answer. When you see yourself in a mirror, light travels from the room lamp to your nose, then from your nose to the mirror, then from the mirror to your eye. The "nose-to-mirror" leg of this journey concerns us here. If your nose travels at the speed of light, and the light leaving your nose travels at the speed of light, does the light ever reach the mirror? Or does it pile up in a little puddle in front of your nose?

This puddle of light seems implausible, but let's admit that the whole situation seems implausible. The speed of light is so large that no railroad coach, no jet plane, no rocket ship has ever moved (with respect to the Earth) at or near the speed of light. The speed of light is

$c = 186\,000$ miles/second or
$c = 670$ million miles/hour.*

(The speed of light is represented by the letter c, from the Latin *celeritas*, speed.) Put in another way, travel from Los Angeles to New York requires

*For non-Americans: The mile is an American unit of length equal to about 1.6 kilometers.

5 hours, by jet;

10 minutes, by space shuttle;

0.015 second, at light speed.

Through reading this book you will discover that many bizarre and unexpected phenomena occur when travel speeds become a substantial fraction of the speed of light—phenomena that fly in the face of all common sense. There is nothing wrong with this: common sense is just a summary of common experience, and we don't commonly experience travel near the speed of light.

This analogy might help: Suppose that one morning I leave my home and travel 1 mile east to buy a carton of milk, then 2 more miles east to pick up a newspaper. What's the shortest route home? For those of us living on Earth it's common sense that the shortest route home is to go west for 3 miles. But if we lived on the equator of a small planet only 4 miles in circumference, then it would be equally obvious that the shortest route home would be to continue east for 1 more mile. Because the Earth is so large, we don't need to consider its curvature in day-to-day navigation. But if we lived on a much smaller planet, then we *would* need to consider its curvature in day-to-day navigation, and this curvature would be common sense to us. My point is that the meaning of "common sense" is not universal but depends on context. In the United States it's common sense to greet friends in the morning with "Hello." In Germany it's common sense to do the same with "Guten Tag."

So let's put the paradox of the mirror into a more familiar context by considering a different version of the nose-to-mirror leg of the light's journey. Suppose that I can throw tennis balls at a speed of 3 miles/hour and that a railroad coach travels at 21 miles/hour. What happens if I throw tennis balls forward while riding in the railroad coach? The answer should be common sense in this familiar context. The tennis balls travel at 3 miles/hour relative to the railroad coach but at 24 miles/hour relative to the Earth.

What does this phrase "relative to" mean? Suppose I'm standing on the sidewalk, looking at a home across the street. A car passes by, moving to the right at 20 miles/hour. Then a bicycle passes by, moving to the right at 5 miles/hour. I would sketch the situation like this:

Earth's frame

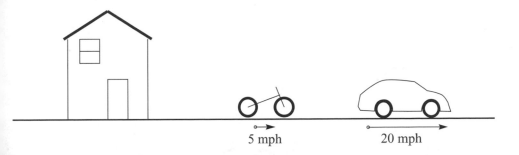

5 mph   20 mph

This sketch shows the speeds relative to the Earth or, as we will often say, "speeds observed from the Earth's reference frame." (We could also say "speeds from my frame" or "speeds from the sidewalk's perspective" or "speeds from the sidewalk's frame" or "speeds in the sidewalk's frame." All of these phrases have the same meaning.)

But I'm not the only person who can observe this scene. The cyclist is just as good a person as I am. From the point of view of the cyclist, the car is still moving, but it's not drawing away from him as quickly as it's drawing away from me; in fact, it's moving at 15 miles/hour as observed from the bicycle's reference frame. Meanwhile, from the point of view of the cyclist, the home is drawing away from the bicycle. In fact, it is moving *left* at 5 miles/hour (or, what is the same thing, right at −5 miles/hour). Here's a sketch of the situation from the cyclist's point of view:

Bicycle's frame

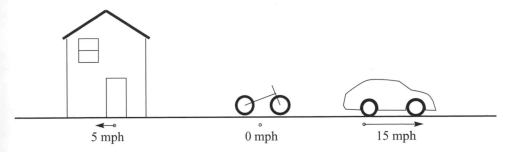

5 mph   0 mph   15 mph

Finally, the same scene can also be examined from the car's point of view, where the situation is

Car's frame

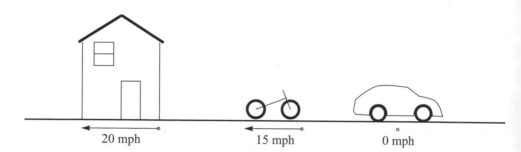

| 20 mph | 15 mph | 0 mph |

Q: Which of these three sketches is right?

A: They're all right. It's customary to measure speeds with respect to the Earth—a sign stating "Speed limit: 65 mph" always means 65 miles/hour with respect to the Earth*—but this is only a custom, not a necessity. You know that there's nothing sacred about the Earth's reference frame: the Earth spins on its axis and orbits the Sun, while the Sun orbits the center of our galaxy, and our galaxy itself rushes toward its neighbor the Andromeda galaxy.

While it's common practice to say "the car moves" rather than "the car moves relative to the Earth," in this book we're going to have to be a lot more careful, and every time we mention motion (or stationarity!) we're going to have to make clear the reference frame we're using. If Tanya and Nathan drive together in a car that moves at 50 miles/hour relative to the Earth, then Tanya moves at 50 miles/hour in the Earth's frame and at 0 miles/hour in Nathan's frame.

In light of our new understanding of the importance of specifying reference frames, we sketch the tennis ball situation twice. Each sketch shows the same situation, but from a different reference frame:

---

*Don't try to get out of a speeding ticket by invoking reference frames: you'll merely irritate the officer who pulled you over.

Earth's frame

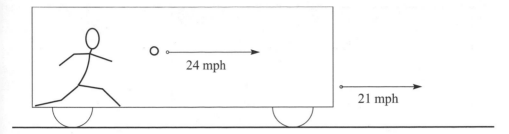

Railroad coach's frame

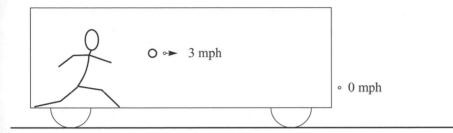

Now the speeds 21 miles/hour (railroad coach relative to the Earth) and 3 miles/hour (tennis ball relative to railroad coach) could have been different numbers—perhaps 57 miles/hour and 9 miles/hour. Let's just name these speeds with the symbols V and v_h, so that

V represents the speed of the railroad coach relative to the Earth, and
v_b represents the speed of the tennis ball relative to the railroad coach.

Using these symbols, the two sketches are

Earth's frame

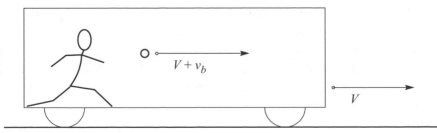

Railroad coach's frame

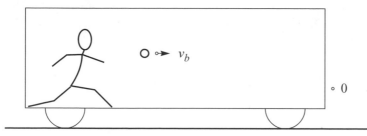

And it's just as clear that the thrown object doesn't have to be a tennis ball: it could have been a bean bag or a burst of light. The paradox of the mirror is exactly the situation sketched above, but with $V = c$ and $v_b = c$, so $V + v_b = 2c$. In short, the paradox is apparently resolved by saying:

> The term "speed of light" is vague. When we say, "The speed of light is $c = 186\,000$ miles/second," we mean that "the speed of light relative to its source is $c = 186\,000$ miles/second." If the source is moving at speed c relative to the Earth, then the light is moving at speed $2c$ relative to the Earth. Light does not "pile up" in front of the passenger's nose, and he sees his face in the mirror as usual.

This resolution of the paradox is complete, satisfying, and wonderfully in accord with our common sense. In addition, it is completely wrong.

PROBLEMS

Everyone knows that you can't learn to swim by watching; you've got to get into the water and do it yourself. It's equally true that you can't learn science by having it "poured into you" as if your mind were an empty flask. You've got to question the observations, criticize the arguments, and test the conclusions. I hope you're doing this throughout the book, but I particularly want you to work on the problems that are presented at the end of most chapters.

If you write up solutions to problems for a course assignment, then don't just give your final numerical answer but show how you obtained that answer. (This is a rule for all intellectual discourse. For example, suppose that in an American literature course you are assigned to write a paper on the influence of Herman Melville. After considerable reading and analysis, you conclude that *Moby Dick* changed the entire landscape of the American novel. If you write an essay consisting only of that single concluding sentence—"*Moby Dick* changed the entire landscape of the American novel"—you'll earn an F for the assignment.) For most of the problems there are hints and skeleton answers in two appendixes. These appendixes do not "give away the solutions" because the solution consists of more than the numerical answer.

The text of this book is a guided tour through the foreign land of relativity. Everyone knows that you can never become intimate with a foreign land through a guided tour. The problems are designed to push you off the guided tour so that you can explore on your own. You will certainly get lost, but you will learn through getting lost. If you try the problems (either on your own or with friends), get lost, and only then use the hints and skeleton answers to help you find your way, then these appendixes will help you learn relativity. If you turn to them immediately without letting yourself get bruised and battered, then they will actually impede your learning.

Thinking about these problems is not a torture designed to keep you indoors on sunny days; it is the most efficient and effective way for you to learn relativity. My promise to you is that I'll never give you a "make-work" problem: Every problem here has a moral—every one is designed to help you learn. Each problem number marked with a ✦ makes a point that will be used later in the book's logical development. I recommend that you work every marked problem and a sampling of the unmarked problems.

1.1 ✦ Speed in various frames. Consider the situation shown on page 5. In addition to the home, the bicycle, and the car shown in the figures, a motorcycle drives to the right at 50 miles/hour relative to the Earth.

    a. In the frame of the bicycle, what is the speed of the motorcycle?

    b. In the frame of the motorcycle, what is the speed of the motorcycle?

1.2 ✦ Distance traveled in various frames. In the situation shown on page 5, find the distance traveled by the car after two hours of travel

    a. as observed from the Earth's frame

    b. as observed from the bicycle's frame

    c. as observed from the car's own frame

REFERENCES

The letter from 16-year-old Albert Einstein to his uncle can be found in

Jagdish Mehra, "Albert Einstein's first scientific paper," *Physikalische Blätter*, 27 (1971) 385–91.
John Stachel, ed., *Collected Papers of Albert Einstein*, vol. 1 (Princeton, NJ: Princeton University Press, 1987), pp. 6–10.

CHAPTER TWO

Space, Time, and Motion

I n the previous chapter we introduced the paradox of the mirror: What would happen if I were moving at the speed of light and looked into a mirror? We proposed a resolution that hinged upon never using the term "speed" by itself, but always using it as part of the phrase "speed with respect to a certain reference frame." Yet at the very end of the chapter, I said that this proposed resolution was wrong.

What's wrong with it? In order to find out, we will first have to look more carefully at such familiar ideas as "speed" and "reference frame."

Speed is defined as

$$\text{speed} = \frac{\text{distance traveled}}{\text{time elapsed during that travel}}.$$

So if a red car travels 60 miles in 1 hour, a blue car travels 120 miles in 2 hours, and a green car travels 1 mile in 1 minute, then each car travels a different distance, and each journey requires a different amount of time, but all three cars travel with the same speed—namely, 60 miles/hour.

> Q: That's not what I would say. I would say that the green car's 1-minute journey is a "fast trip," while the blue car's 2-hour journey is "slow and boring."
>
> A: The word *speed*, like most words, has several meanings. (The *Oxford English Dictionary* lists 32 meanings for *speed* as a noun—and it's a verb, too!) You're using the meaning relating to "amount of time," but throughout this book, I will use the meaning relating to "amount of distance traveled in a standard time" (such as 1 hour). It is in this sense that all three journeys have the same speed.

The idea of "reference frame" might sound sophisticated or esoteric, but it's just the familiar idea, advanced in the previous chapter, that an object's speed depends on the platform from which that speed is measured.

For example, suppose a ladybug takes 2 seconds to fly from the back window of a car to the windshield 6 feet forward,* and does so while the car is moving down the street at 30 feet/second (about 20 miles/hour). Then, in the car's reference frame the speed of the ladybug is

$$\text{speed} = \frac{\text{distance traveled}}{\text{time elapsed}} = \frac{6 \text{ ft}}{2 \text{ s}} = 3 \text{ ft/s}.$$

To find the speed in the Earth's reference frame, recognize first that the flight of the ladybug still requires 2 seconds, but also that during those 2 seconds the car moves 60 feet down the street. So relative to the street, the ladybug has traveled a distance of 66 feet and its speed is

$$\text{speed} = \frac{\text{distance traveled}}{\text{time elapsed}} = \frac{66 \text{ ft}}{2 \text{ s}} = 33 \text{ ft/s}.$$

Car's frame

$$\rightarrow | \ 6 \text{ ft} \ | \leftarrow$$

ladybug leaves
rear window

ladybug reaches
windshield
2 seconds later

*For non-Americans: The foot is an American unit of length equal to about 30 centimeters. As the name suggests, it is about the length of a typical human adult's foot.

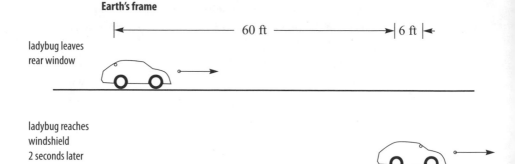

Earth's frame

|← —————————— 60 ft —————————— →| 6 ft |←

ladybug leaves
rear window

ladybug reaches
windshield
2 seconds later

You know that a statement such as "Philadelphia is 300 miles away" is incomplete because the statement is correct when uttered in Boston but not when uttered in San Francisco. A complete statement is something like "Philadelphia is 300 miles away from Boston." In exactly the same way, a statement such as "The ladybug's speed is 3 ft/s" is incomplete. To make the statement complete you must instead say, "The ladybug's speed is 3 ft/s relative to the car" or "The ladybug's speed is 3 ft/s in the car's reference frame."

More generally, if

v_b represents the speed of the bug in the car's frame,
V represents the speed of the car relative to the Earth, and
w_b represents the speed of the bug in the Earth's frame,

then

$$w_b = v_b + V.$$

The idea of reference frame comes up also in air travel. Suppose you're flying in a jetliner, and the captain announces: "We have reached our cruising speed of 500 miles/hour. The air is smooth so I'll turn off the 'fasten seat belt' sign, and the flight attendants will serve coffee." You know that the captain means "500 miles/hour" relative to the Earth below. When a steward pours coffee, the coffee falls smoothly from its pot to the cup waiting below it. The coffee does not leave the pot, realize that it's moving at 500 miles/hour relative to the Earth, and somehow start its fall with zero speed relative to the Earth. If it did, then there would be coffee stains all over the back wall of every jetliner cabin! (Not to mention that it would be dangerous to walk in the jetliner aisle while coffee was being served, because slugs of coffee would fly past you at 500 miles/hour toward the back of the cabin.)

Furthermore: If you were to release a clutch of butterflies into the jetliner cabin, they would fly fore and aft with equal ease, and would not have to flap their wings mightily to keep up with the 500 mile/hour jetliner.

These observations, and many more like them, convince us that

If one reference frame moves uniformly relative to another, then the two are equally good frames for observing nature, and two identical internal experiments performed in the two frames will give identical results.

(By an "internal experiment," I mean one conducted using only objects embedded within the frame, such as pouring coffee into a cup. This excludes experiments like "look out the window and see whether you're moving" or "roll down the window and see whether there's a breeze.") This is called the *principle of relativity*, and it's an old idea, first articulated clearly by Galileo in 1632.

Perhaps you've encountered the following situation, which is a good test of the relativity principle. You're seated in a subway train, stopped at a station and waiting to go north. On the adjacent track a train is stopped and waiting to go south. Suddenly you get the impression that your train is moving, but you're startled because you haven't felt the jerk of the brakes being released. After a moment's confusion, you understand what's going on: Your train is still stopped at the station, but the southbound train has started to move. This experience, like the coffee-pouring experiment, is a qualitative test of the principle of relativity, which has also been tested quantitatively to high accuracy numerous times.

Notice that the principle of relativity involves two reference frames moving past each other "uniformly." If the motion is *not* uniform, if it involves speeding up or slowing down or changing direction, then the two frames *don't* yield the same results for the same experiments. For example, if you're in a jetliner accelerating down the runway, moving faster and faster, then you're tossed back against your seat in a way that you aren't if the jetliner's at rest on the ground or moving at a constant speed in the air. If you try to pour coffee while your jetliner's taking off, the coffee will not pour into the cup below it but will instead end up on your shirt.

Similarly, if you hang a pendant from your car's rearview mirror, then that pendant hangs straight down while the car moves with constant speed relative to the Earth (including zero speed). But if the driver accelerates, the pendant swings back (relative to the car). If the driver brakes, the pendant swings forward. And if the driver turns left, the pendant swings right. All these experiments show that if two reference frames *aren't* moving uniformly relative to each other, then the two are *not* equally good frames for observing nature, and two identical internal experiments performed in the two frames will *not* give identical results.

Any reference frame that passes the "coffee-pouring test" is called an *inertial reference frame*. The technical requirement is that an inertial frame is "one in which any body moves in a straight line at constant speed unless acted upon by a force," but rather than get caught up in complicated words it's best to think of a concrete experiment, such as the coffee-pouring test.

Q: The Earth is rotating, so a reference frame attached to the Earth's surface is moving in a curved line, so it's not inertial!

A: That's right: the Earth's surface is not exactly an inertial frame. Because the Earth's radius is so large, this effect is small enough that it escapes the attention of all but the most diligent experiments. The Earth's surface is not a perfect inertial frame, but it's a good approximation.

In the rest of this book we'll take advantage of this approximation and speak of the "Earth's reference frame" as if it were inertial. If this disturbs you, then whenever I write "the Earth's frame," substitute the phrase "a perfect inertial reference frame."

In most cars, you can tell whether you're stationary or moving relative to the ground, without looking out the window, by feeling the vibrations of the car. That's because a car moving down a street hits tiny bumps in the pavement and thus changes its up and down motion frequently and unpredictably. The car's motion is not in a perfect straight line, so the car's reference frame is not perfectly inertial.

Einstein's theory regarding inertial reference frames, which he discovered in 1905, is called *special relativity*. The first 16 chapters of this book concern special relativity. These chapters discuss objects and reference frames in uniform motion, without worrying about how they got into motion. Einstein's theory regarding noninertial reference frames, which he discovered in 1915, is called *general relativity*. Chapters 17 through 19 introduce general relativity.

We've gotten a lot done in this chapter: We've clarified the notions of "speed" and "reference frame," we've established that any specification of speed must be with respect to a particular reference frame, and we've put forward the principle of relativity. On the other hand, we haven't directly addressed the paradox of the mirror. So let's ask again: If light leaves my nose at speed c relative to my nose, and my nose is traveling at speed V relative to the Earth, at what speed does the light move relative to the Earth? The answer $V + c$ seems obvious.

PROBLEMS

2.1 ✦ Backseat shooting. You are driving a car due east down a highway. Your young daughter is strapped into her child seat in the back, on the south side of the car, holding a loaded suction dart gun. She decides to shoot a suction dart onto the opposite back window. She fires the suction dart at (*a*), and the dart sticks on to the window a moment later at (*b*). The situation as observed from the car's frame is shown below. As you can see, the suction dart flies straight north.

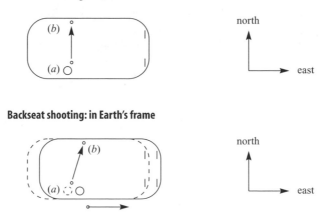

Backseat shooting: in car's frame

Backseat shooting: in Earth's frame

But in the Earth's reference frame, the car moves between the firing of the dart and its sticking onto the window. The situation at firing is shown with dashed lines and the situation at sticking is shown with solid lines. As you can see, the dart flies in a direction somewhat east of north.

     a. If the speed of the car relative to the Earth is the same as the speed of the dart relative to the car, in what direction does the dart travel in the Earth's frame? (North? East? Northeast? Between north and northeast? Between northeast and east?) What if the speed of the car relative to the Earth is less than the speed of the dart relative to the car?

     b. Argue that the dart travels faster in the Earth's frame than it does in the car's frame.

REFERENCES

The principle of relativity was first clearly articulated in

Galileo Galilei, *Dialogue Concerning the Two Chief World Systems* (Florence, 1632),

a book still worth reading today both for its clarity and for its humor. I recommend the 2001 Modern Library paperback edition, translated by Stillman Drake, where the principle of relativity appears in a shipboard story on pp. 216–18.

Early quantitative tests of the relativity principle were performed by Frederick Thomas Trouton and Henry R. Noble:

F. T. Trouton, "Effect on charged condenser of motion through the ether," *Transactions of the Royal Dublin Society*, 7 (1902) 379–84.

F. T. Trouton and H. R. Noble, "Forces acting on a charged condenser moving through space," *Proceedings of the Royal Society of London*, 72 (1903) 132–33.

Testing has continued with ever higher accuracy and in an ever greater variety of circumstances. For example:

J. A. Lipa, J. A. Nissen, S. Wang, D. A. Stricker, and D. Avaloff, "New limit on signals of Lorentz violation in electrodynamics," *Physical Review Letters*, 90 (14 February 2003) 060403/1–4.

B. R. Heckel, C. E. Cramer, T. S. Cook, E. G. Adelberger, S. Schlamminger, and U. Schmidt, "New *CP*-violation and preferred-frame tests with polarized electrons," *Physical Review Letters*, 97 (14 July 2006) 021603/1–4.

It would be very exciting indeed if the principle of relativity were proven incorrect—a real Nobel Prize–winning experiment—and therefore more experimental tests are in the works.

CHAPTER THREE

The Strange Behavior of Light

In science, the test of an assertion is not whether it is obvious, or satisfying, or wonderfully in accord with common sense. The test of an assertion is whether or not it agrees with what actually happens in nature.

The assertion that "the speed of light is c relative to its source, so if the source moves at speed V relative to the Earth, then the emitted light moves at speed $V + c$ relative to the Earth" has been tested many times, and these tests have shown it to be *false*. Instead, these tests have found that

> *The speed of light is the same in all inertial reference frames, regardless of the speed of the source.*

One of the most impressive of these tests was performed in 1964 at CERN, the European Organization for Nuclear Research. (The acronym derives from the laboratory's former French-language name, Conseil Européen pour la Recherche Nucléaire.) As a light source this test used a particle called a neutral pion. You don't need to know anything about neutral pions except that they are sources of light, like flashlights (or noses). If you put a neutral pion on the table in front of you, it will soon emit a burst of light moving at speed c.

But neutral pions are much less massive than flashlights, so it's relatively easy to get them moving at high speeds. In the CERN experiment, neutral pions were measured moving (relative to the Earth) at $0.99975\,c$ before they emitted their burst of light. Yet the light departing from those neutral pions was measured to move (relative to the Earth) at speed c, not at $1.99975\,c$.

This is not the only test of the claim that the speed of light is the same in all inertial reference frames. One early test was performed by Albert A. Michelson in 1881. Michelson knew that the Earth rotated on its axis and orbited the Sun, so he suspected that light launched in the direction of the Earth's motion (say east-west) would travel at a speed different from that of light launched in the perpendicular direction (north-south). Michelson designed an experiment to measure this difference. It involved splitting a light beam into two pieces and sending one piece on a journey, say, north to a mirror and then back south again, while the other piece journeyed east to a mirror and then back west again. The two light beams were then recombined, and (using an ingenious technique relying on the wave character of light) the difference between the north-south transit time and the east-west transit time could be determined.*

This was a difficult experiment to perform. For example, the apparatus had to be able to rotate easily so that it could change orientation as the direction of Earth's motion changed through daily rotation and yearly revolution. At the same time the positions of the mirrors and the light sources had to be rigidly fixed while any single observation was being made: because of the ingenious detection technique mentioned above, a vibration or expansion that moved a mirror as little as one wavelength of light would invalidate the entire observation. And of course Michelson was looking for a very small effect to begin with: the orbital speed of the Earth is 20 miles/second, which is large by human standards but tiny compared to the light speed of 186 000 miles/second.

Nevertheless by 1881 Michelson had refined his apparatus to the point where it could measure a difference in speeds as small as 13 miles/second. Yet no difference could be detected.

In 1887 Michelson (this time with collaborator Edward Morley) repeated the experiment with a more sensitive apparatus, capable of measuring a speed difference as small as 3 miles/second. Still no difference could be detected.

This experiment has been tried again and again over the years. In 2003 Holger Müller and his collaborators repeated the experiment with a panoply of high-tech equipment: lasers, servomotors, and cryogenic optical resonators made of crystalline sapphire. This apparatus could measure a speed difference as small as 0.4 *billionth* of a mile/second. Yet again no difference could be detected.

---

*A particularly clever aspect of this experiment is that it measured the *difference* in two transit times without measuring the two transit times themselves.

Q: Do you mean to say that in 122 years of repeating the Michelson-Morley experiment, not once has a speed difference been detected?

A: In experiments performed from 1902 to 1933, Dayton Miller detected a difference. It wasn't 20 miles/second, but it did vary with an annual pattern, as you would expect if it depended on the Earth's orbit around the Sun. At first no one (including Miller) could understand why Miller got one result and everyone else got a different result. But in 1955 R. S. Shankland and collaborators reanalyzed Miller's data* and ascribed the difference to an experimental flaw: The signal detected depended on temperature (which of course also varies with an annual pattern). Apparently Miller's results were due to one arm of his apparatus expanding more than the other. An even more sophisticated reanalysis was performed in 2006 by Thomas J. Roberts, who found that Miller's data were "not statistically significant."

Q: Was Miller ostracized by the scientific establishment for questioning the Einstein orthodoxy?

A: Hardly. In 1921 he was elected to the U.S. National Academy of Sciences, a signal honor. In that same year Einstein visited Miller and encouraged him to continue his measurements. Soon after this visit Einstein wrote that if Miller's results were confirmed, "the whole relativity theory collapses like a house of cards," and Miller wrote that Einstein was "not at all insistent upon the theory of relativity." In 1927 Ambrose Swasey donated $100 000 to the Case School of Applied Science to support Miller's work. In 1924, 1927, 1933, and 1936 Miller was awarded honorary degrees.

These are not the only tests of the proposition that the speed of light is the same in all inertial reference frames. In 1997 Yuan Zhong Zhang surveyed the field and listed 32 experiments, with light sources ranging from sodium lamps to stars to the Sun to mercury arc lamps to lasers to pions, and with several different techniques for measuring light speed, but all of them showed the speed of light to be the same in all inertial frames.

Q: I read that light moves slower in air than in vacuum, and slower in glass than in air. In fact, I saw in the newspaper that someone had created a material in which the speed of light was 38 miles/hour! How can the speed of light be "the same" if it varies depending on the medium?

*Miller had requested this before his 1941 death.

A: When light moves through a medium, it travels through space for a while and then strikes an atom, which absorbs it with an increase in its own energy. After a while the atom releases some light and goes back to its original energy status. This newly released light travels through space until it strikes and is absorbed by another atom, and there the process starts over again. While the light is actually moving through space, between atoms, it travels at speed c. But some of the time the light is not moving at all—it's absorbed by an atom. The net result is that, on average, the light travels slower than c. In the same way a car that can move at 60 miles/hour travels somewhat less than 600 miles in 10 hours because the driver takes rest stops. (This discussion is not 100% accurate because it ignores the quantum character of light, but it gives the right impression and does so in a single paragraph. To produce a 100% accurate discussion would double the size of this book.)

While the light is actually traveling—between atoms, as it were—it travels at speed c, the speed of light in vacuum. This is the speed that's the same in all reference frames. The average speed—taking into account the "rest stops"—is less than c, and this average speed is *not* the same in all reference frames. If you think that my language is too loose, then every time I write "speed of light" you should read "speed of light in vacuum."

Q: Why do you use units like miles and feet and hours? In elementary school they told me that science is done using metric measurements.

A: Science is about nature, careful observation, and free and disciplined reasoning, not about social criteria like metric measurements or neckties or white lab coats.

Q: Your claim is ridiculous! We know that the speed of a ladybug depends on reference frame, the speed of a tennis ball depends on reference frame, and the speed of a bullet depends on reference frame. How can light be any different?

A: The claim is certainly not in accord with common sense, but it *is* in accord with experiment, and that's what matters.

It's the job of the rest of this book to show how this fact can be true. But I'll give you one hint right now: We'll find out that it's not light that's special, but the speed of light. Anything that moves at the speed of light—light, X rays, gamma rays, infrared radiation, radio waves, gravity waves—moves with the same speed in all inertial reference frames.

In fact, we'll find that the commonsense result $w_b = v_b + V$ (from page 12) is never *exactly* correct, although it's an excellent approximation when the speeds V and v_b are much less than light speed. If you apply this formula to higher and higher speeds, you'll find that it gives worse and worse results until, when

v_b represents not the speed of a ladybug but the speed of a burst of light, you'll find that the correct formula is $w_b = v_b$: in other words, that the speed is the same in both reference frames.

It will be an exciting step in our intellectual journey when we produce the equation that correctly interpolates between the commonsense behavior for low speeds on one hand and the "same in all frames" behavior for light speed on the other hand. (If you can't wait, you can peek ahead and find that equation on page 100.)

Q: The question I really want answered is not "How can the speed of light be the same in all reference frames?" but "Why is the speed of light the same in all reference frames?"

A: I don't know the answer to that question, and I don't even know how to approach it. Questions beginning in "why" are rarely scientific questions. It's hard to think of an experiment or observation that would answer such a question. For example: "Why is our universe three-dimensional?" I don't know—but experiment shows that it is! The three-dimensional character of our universe is a fact that is unexplained but familiar, so most people don't bother asking why it's true. The constancy of the speed of light is just as unexplained, but it's unfamiliar, so you're more likely to wonder about it.* The job of this book is to make you familiar with the real, relativistic universe that we live in, not with the commonsense, low-speed universe that we think we live in.

In the year 1633, when the Holy Office of the Inquisition imprisoned Galileo for holding that the Earth orbited the Sun, the inquisitors were trying to get rid of unfamiliar and unpleasant facts by forcing them to go away. Today, we realize that it's our job to listen to nature, not to dictate to nature. Don't make the inquisitors' mistake.

Q: But I'm a "why" sort of person. I care a lot more about the "why" than the "how."

A: I'm not saying that "why" questions are unimportant—on the contrary, deep questions like "Why is life worth living?" and shallow questions like "Why should I eat broccoli?" are central to our daily lives. However, they're not scientific questions, and this is a book about science, so I won't try to answer them.

But even if you're more concerned with "why" questions than "how" questions, it's still worth your time to uncover the "how" before moving on to the "why." First, the scientific "how" questions are a lot easier to answer than the beyond-science

---

*When he was 5 years old my son Colin, frustrated with the demands of being a younger brother, asked me, "Why is Greg two years older than me?" The answer—"He was born two years earlier"—didn't help because it merely restated the question.

"why" questions. Second, the "how" answers lay the foundation for the "why" questions.

For example, in the year 1600 some people were interested in a deep question: "Why do the planets move in circular fashion?" Numerous explanations were advanced, some involving crystalline spheres that carried the planets. There was even speculation about the sound that two such spheres made as they rubbed against each other: "the music of the spheres."

In that same year, others were interested in the question: "How can one predict the position of planet Mars?" This question is far more mundane than celestial speculations about spheres, but it was through pursuing this question that Johannes Kepler discovered that planets moved in ellipses, not circles, and hence disposed of the entire question of spheres. The deep "why" questions had all been built on a faulty foundation, a misunderstanding of *how* planets move.

Units

Most of the situations in this book will involve ordinary, human-sized distances. Typically we will measure these distances using as a unit the foot.

Because the speed of light is so large (compared with typical human speeds) most of the situations in this book involve very small times. For example, a ball traveling at half the speed of light moves 1 foot in only 2.034×10^{-9} second. It becomes very tiresome to say the names of such short time intervals: over and over again, you'll have to pronounce "ten to the negative nine second." So I'll introduce a new unit of time, a very short unit, named the "nan." A nan is the amount of time that it takes light to travel exactly 1 foot, so it is a very short time by human standards: 1 nan is only 1.017×10^{-9} second. The name "nan" comes from the fact that a nan is nearly equal to a "nanosecond," which is the name for a billionth of a second (10^{-9} second). In these units, the speed of light is exactly $c = 1$ foot/nan, so a ball traveling at half the speed of light moves one foot in two nans.

PROBLEM

3.1 ✦ **Looking at a clock.** Cynthia stands 42 feet from a clock and sees it reading 431 nans. What time is it?

REFERENCES

The CERN neutral pion experiment is described in

T. Alväger, F. J. M. Farley, J. Kjellman, and I. Wallin, "Test of the second postulate of special relativity in the GeV region," *Physics Letters*, 12 (1964) 260–62.
T. Alväger, J. M. Bailey, F. J. M. Farley, J. Kjellman, and I. Wallin, "The velocity of high-energy gamma rays," *Arkiv för Fysik*, 31 (1966) 145–57.

The various versions of the Michelson-Morley experiment described are

A. A. Michelson, "The relative motion of the Earth and the luminiferous aether," *American Journal of Science*, 22 (1881) 120–29.
A. A. Michelson and E. W. Morley, "On the relative motion of the Earth and the luminiferous ether," *American Journal of Science*, 34 (1887) 333–45.
H. Müller, S. Herrmann, C. Braxmaier, S. Schiller, and A. Peters, "Modern Michelson-Morley experiment using cryogenic optical resonators," *Physical Review Letters*, 91 (11 July 2003) 020401/1–4.
Dayton C. Miller, "The ether-drift experiment and the determination of the absolute motion of the Earth," *Reviews of Modern Physics*, 5 (1933) 203–42.
R. S. Shankland, S. W. McCuskey, F. C. Leone, and G. Kuerti, "New analysis of the interferometer observations of Dayton C. Miller," *Reviews of Modern Physics*, 27 (1955) 167–78.
Thomas J. Roberts, "An Explanation of Dayton Miller's Anomalous 'Ether Drift' Result," http://arxiv.org/abs/physics/0608238.

The reactions of Einstein and Miller to their meeting are described in

Ronald W. Clark, *Einstein: The Life and Times* (New York: World, 1971), p. 328.
Loyd S. Swenson, *The Ethereal Aether* (Austin: University of Texas Press, 1972), p. 195.

The 32 experiments on the constancy of the speed of light are listed and described in

Y. Z. Zhang, *Special Relativity and Its Experimental Foundations* (Singapore: World Scientific, 1997).

The experiments on "slow light" by Lene Vestergaard Hau and her colleagues were reported in

L. V. Hau, S. E. Harris, Z. Dutton, and C. H. Behroozi, "Light speed reduction to 17 metres per second in an ultracold atomic gas," *Nature*, 397 (18 February 1999) 594–98.

and popularized through articles in, for example, the *New York Times* on 18 February 1999 and 30 March 1999.

PART II / Uncovering Relativity

CHAPTER FOUR

Time Dilation

Sometimes we see time adorned with green foliage as pleasant as an
angel; And then suddenly he changes and becomes very strange.
—FROM THE FRENCH TAPESTRY "TIME" (C. 1505)

The speed of light does not behave in the commonsense way that we expect. But
speed is defined as

$$\text{speed} = \frac{\text{distance traveled}}{\text{time elapsed}}.$$

So if speed doesn't behave in the commonsense way then it must be because (*a*) time
doesn't behave in the commonsense way, or (*b*) distance doesn't behave in the
commonsense way, or (*c*) both. We begin with time and specifically with clocks.

There are a lot of clocks in the universe, and a lot of different kinds of clocks. A
handsome grandfather clock works through the oscillation of a pendulum. The watch
on my wrist works through the oscillation of a quartz crystal, which is processed by
some electronics to make a digital readout. Older wristwatches work through the
oscillation of a balance wheel by tightly coiled springs, which is processed through
an ingenious gear mechanism to turn hands on the watch's face.

Because we spend so much time looking at the display of a clock (the digital
readout or the face and hands), it's tempting to think that this is the clock's core
element. No. The core element of a clock is the oscillating pendulum, or the quartz
crystal, or the balance wheel—whatever element it is that changes with time. The
display merely serves to count the number of oscillations of the core element.

Indeed, some clocks have neither a digital readout nor a face and hands. The Sun
in the sky is a clock, marking off 24 hours as it goes from its highest position in the
sky on Monday to its highest position in the sky on Tuesday. In the first century BC,
Athenians kept track of time at the Tower of the Winds, where water dripped from a

tank at a nearly constant rate. A waterfall, slowly and tirelessly wearing away rock, is a clock: you can find the age of a waterfall by measuring the amount of rock it has worn away. Your own body is loaded with clocks: your heart beats about once each second; your fingernails and hair grow, marking the passage of time;* and your brain maintains a good sense of the difference between a brief lecture and a lengthy one, even if you don't glance at your watch. Each of these is a clock. These biological clocks are not the most accurate of all clocks (your heart beats faster when you run), but they are clocks nevertheless.

The most accurate kind of clock known today is the fountain atomic clock. There are two such clocks—one in Paris and one in Boulder, Colorado—and they are so accurate that if they ran for 20 million years, they would drift apart from each other by only 1 second.

These various clocks differ in mechanism, accuracy, convenience, durability, and expense, but they have in common that they all tell time. This, indeed, is the meaning of the word *time*: it's the property held in common by all clocks.†

I'm going to introduce you to a kind of clock called a light clock. As you will soon see, it's not a convenient clock for strapping to your wrist, but it's very convenient for uncovering the properties of time in special relativity.

The Light Clock

The workings of a light clock (as observed from its own reference frame) are sketched below. It starts when a strobe lamp flashes briefly at (a). The burst of light emitted goes up at speed c and reflects off a mirror at (b). Then the burst goes down at speed c and is absorbed by a light detector at (c). When the detector absorbs the burst, it does two things: First, it increments the digital display from 000 to 001. Second, it triggers the strobe lamp to flash again. A second burst of light goes up, then down, and once it is absorbed, the digital display increments from 001 to 002 and the strobe lamp flashes once again. The process repeats over and over. This is the repetition— analogous to the oscillation of a pendulum or a quartz crystal or a balance wheel or a heart—at the core of the light clock.

---

*When my moustache tickles my upper lip, I know that it has been 6 weeks since I last trimmed it.
†Most definitions define words and concepts in terms of other words and concepts. Attempts to define *time* in this way invariably lead to circularity. (For example, my dictionary defines *time* to mean "duration," but defines *duration* to mean "continuance in time.") To avoid this problem, we define *time* here not in terms of words but in terms of actions performed in nature: specifically, in terms of how to build a clock. This kind of definition is called an *operational definition*.

Light clock observed from its own frame

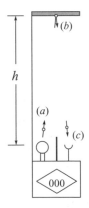

Just as the amount of time elapsed during one swing of a pendulum depends on the pendulum's length, so the amount of time elapsed during one cycle of the light clock depends on the light clock's height. If you want your light clock to tick off nans, then it should be half a foot high. If you want your light clock to tick off seconds, then it should be 93,000 miles high. In general, if the height of the light clock (from strobe lamp to mirror) is h, then the time for one cycle of the light clock is $2h/c$.

You might object that the burst of light travels not merely up and down, but also a little bit from the left to the right, in order to move from the strobe lamp over to the light detector. Thus, the cycle time is just a shade more than $2h/c$. In this you'd be absolutely correct. To keep our analysis straightforward, we can design and build

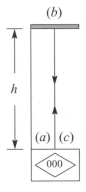

Fig. 4.1. Light clock observed from its own reference frame.

very small strobe lamps and light detectors, and enclose them within the box, making the left-to-right motion negligible. Our improved light clock is sketched in figure 4.1.

Light Clock in Motion

Above, we described the light clock observed from the light clock's own frame. What happens if we analyze the light clock from a different frame, a frame in which the light clock moves? (We can call this either "a light clock moving to the right" or "an observer moving left past a light clock." According to the principle of relativity, these two situations are identical.)

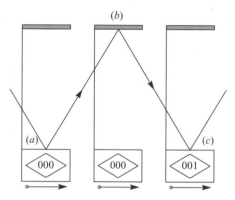

Fig. 4.2. Light clock observed from a reference frame in which it moves.

What is going on here? This figure is *not* a snapshot: instead, it shows the light clock at three different instants and traces the path of light between those instants. At (*a*) the strobe lamp generates a burst of light. Some time later, at (*b*), the burst is reflected from the mirror. The clock moves some distance rightward while the light travels upward, so the figure shows (*b*) to the right of (*a*).* Later still, at (*c*), the burst is detected, the counter increments, and the strobe lamp flashes once more to start the next cycle. Again, the clock has moved some distance rightward while the light travels downward. The previous two figures show exactly the same actions: they differ only in reference frame—in the platform from which they observe those actions.

---

*If you're puzzled to find a burst of light going straight up and down when observed from the light clock's frame, but the same burst moving at an angle here, then go back to problem 2.1, "Backseat shooting," on page 15 to see this same phenomenon in a more familiar context.

How much time does the process (a) to (b) to (c) require? Suppose, for concreteness, that our light clock is half a foot high. Then, *as observed from the light clock's frame* (see fig. 4.1), the light moves half a foot up between (a) and (b), then half a foot down between (b) and (c). From (a) to (b) to (c) the light has traveled 1 foot. The light travels at speed $c = 1$ foot/nan in this reference frame. Thus, one nan has elapsed, and the clock ticks off one nan.

But *as observed from a frame in which the light clock moves* (see fig. 4.2), the light moves both up *and to the right* between (a) and (b), then both down *and to the right* between (b) and (c). From (a) to (b) to (c) the light has traveled *more than* 1 foot. The light travels at speed $c = 1$ foot/nan in this reference frame too, so the longer distance requires a greater time. Thus, *more than* 1 nan has elapsed, yet still the clock ticks off only 1 nan!

In the light clock's own frame, 1 nan has elapsed and the clock ticks off 1 nan. In the frame in which the light clock moves, more than 1 nan has elapsed, yet the clock ticks off 1 nan. We conclude that *a moving light clock ticks slowly.*

Is this a defect in the light clock? No. Suppose we have a pendulum clock and a light clock side by side, and set them into motion (relative to, say, the Earth). The light clock now ticks slowly. If the pendulum clock *doesn't* tick slowly, then we have an internal experiment that allows us to state whether the two clocks are "moving" or "stationary." (If the two clocks remain in synch, then the clocks are stationary. If the light clock falls behind the pendulum clock, then the two clocks are moving.) The principle of relativity (the "coffee-pouring principle") tells us that this is impossible. Both clocks have to tick slowly. This slowing does not reflect the construction of any particular type of clock; instead, it reflects the character of time. We have uncovered that

A moving clock ticks slowly.

This principle is called *time dilation.*

Just How Slowly Does the Moving Clock Tick?

To make the situation concrete, consider a light clock carried down the street in a car. The light clock is stationary in the car's frame, but moving at speed V in the Earth's frame.

In the car's frame, a certain time elapses between ticks. We'll call this time T_0. In the Earth's frame, a longer time elapses between ticks, but the clock still ticks off

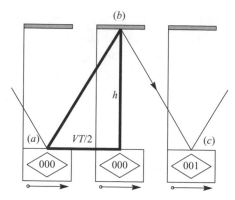

Fig. 4.3. Light clock observed from a reference frame in which it moves. The heavy triangle has height h, base $VT/2$.

only time T_0. We'll call this longer time T. This section finds a formula for T in terms of T_0 and V.

Problem 1.2, "Distance traveled in various frames," on page 9, shows that the distance traveled by a car is different in different reference frames. In the same way, the distance traveled by the burst of light between two ticks of the light clock is different in different reference frames. Once we find those two different distances, we'll be able to relate the two different times.

Consider a light clock of height h. In the car's frame, the light moves a distance $2h$ between ticks.

In the Earth's frame, the light moves a longer distance between ticks. Figure 4.3 is the same as figure 4.2, except that I have traced in a triangle using heavy lines. The heavy diagonal is half the distance that the light travels between ticks, and this is the distance we need to find. The height of the triangle is, of course, h. The base of the triangle is the distance that the light clock moves during half the time between ticks, or

$$\text{base} = \text{speed} \times \text{time} = V \times \frac{T}{2}.$$

Do you remember the Pythagorean theorem of geometry? This relates the lengths of three sides of a right triangle—the diagonal, the height, and the base—stating that

$$\text{diagonal}^2 = \text{height}^2 + \text{base}^2.$$

Applied to the triangle in the figure, the Pythagorean theorem tells us that

$$\text{diagonal}^2 = h^2 + (VT/2)^2,$$

so the diagonal has length $\sqrt{h^2 + (VT/2)^2}$, and the light travels a distance $2\sqrt{h^2 + (VT/2)^2}$ between ticks. Now that we have the two distances, we can relate the two times.

In the car's frame, one tick requires time T_0, and it happens after light has moved a distance $2h$. This time and distance are related through

distance traveled = speed × time elapsed

$$2h = cT_0. \tag{4.1}$$

In the Earth's frame, one tick requires time T, and it happens after light has moved a distance $2\sqrt{h^2 + (VT/2)^2}$. This time and distance are related through

distance traveled = speed × time elapsed

$$2\sqrt{h^2 + (VT/2)^2} = cT. \tag{4.2}$$

To relate T and T_0, put equations (4.1) and (4.2) together and eliminate h between the two of them. This is just algebra. Start with equation (4.2):

$2\sqrt{h^2 + (VT/2)^2} = cT$	Plug in equation (4.1) for h
$2\sqrt{(cT_0/2)^2 + (VT/2)^2} = cT$	Divide both sides by 2
$\sqrt{(cT_0/2)^2 + (VT/2)^2} = cT/2$	Square both sides
$(cT_0/2)^2 + (VT/2)^2 = (cT/2)^2$	Multiply both sides by 2^2
$(cT_0)^2 + (VT)^2 = (cT)^2$	Divide both sides by c^2
$T_0^2 + (V/c)^2T^2 = T^2$	Solve for T_0^2
$T_0^2 = (1 - (V/c)^2)T^2$	Take the square root of both sides
$T_0 = \sqrt{1 - (V/c)^2}\,T$	Solve for T
$T = \dfrac{T_0}{\sqrt{1 - (V/c)^2}}.$	

And now we have it: the quantitative relation between T_0 (the time elapsed between two clock ticks in the clock's frame) and T (the time elapsed between two clock ticks in a frame in which the clock moves at speed V).

Q: Ugh! My brain is reeling! I haven't seen so many symbols since I visited the Vatican!

A: If your brain is reeling because it's discovering new and unexpected properties of time, then that's a good thing. Anyone who *doesn't* find time dilation strange and discomforting hasn't thought about it enough, so your discomfort is a sign that you're thinking.

If your brain is reeling because you haven't exercised using algebra for a long time, then focus on the qualitative discovery that T is *larger* than T_0—that a moving clock ticks *slowly*. The light clock, the symbols, the algebra—these are just scaffolding to propel the argument. The result is that time behaves differently in different frames, and time existed (and dilated!) long before humans were around to build light clocks and write symbols.

Q: Do you expect me to memorize this derivation?

A: No. The derivations in this book are important because they show a logical connection between experiment and conclusion, but they are not the only way to make these connections. For example, when Einstein discovered relativity in 1905, he used a very different path from the one that we're taking.* Instead of memorizing derivations, learn how to work with these concepts by doing a few problems.

Q: If a moving clock ticks slowly, why haven't I ever noticed it before?

A: The effect is minute at everyday speeds. For example, the space shuttle orbits Earth at a speed of about 5 miles/second. A clock moving past us at that speed will lose time relative to our clocks. But how much time? The time dilation formula tells us that while our clocks tick off 1 hour (T) the moving shuttle clock will tick off 1.3 millionth of a second less than an hour (T_0). It is a rare clock that can measure time with this sort of accuracy!

As the speed of the clock increases, the effect becomes more noticeable. While our clocks tick off 1 hour,

a clock moving at half the speed of light ticks off 52 minutes,

a clock moving at 3/5 the speed of light ticks off 48 minutes,

a clock moving at 4/5 the speed of light ticks off 36 minutes,

a clock moving at 99% the speed of light ticks off 8.5 minutes.

Q: Why stop there? What if the clock moves at the speed of light? At twice the speed of light?

A: The time dilation formula is $T = T_0/\sqrt{1 - (V/c)^2}$. If $V = c$, then $\sqrt{1 - (V/c)^2} = 0$, so this formula involves division by zero. It's hard to understand how to interpret this. If $V = 2c$, then $\sqrt{1 - (V/c)^2} = \sqrt{-3}$, so the formula involves taking the square root of a negative number. It's even harder to understand how to interpret this!

---

*Why don't I guide you along Einstein's path? Because it uses even more symbols and algebra!

Fortunately, we will never need to interpret these absurdities. We will discover in chapter 10, "Speed Limits," that no clock can travel at or above the speed of light.

Q: I see you using the formula, but it just doesn't make sense to me.

A: Is there any reason why it should make sense? Our brains evolved for the purpose of finding food and avoiding being eaten, not for understanding relativity! If we want to explore this brave new world, we are going to have to push our brains into unfamiliar territory where they are not comfortable.

Albert Einstein made this point by saying, "Common sense is nothing more than a deposit of prejudices laid down in the mind before you reach eighteen." But Henry David Thoreau helpfully pointed out that "it is never too late to give up our prejudices."

Q: All you've done so far is talk. Has anyone ever observed time dilation? I'm suspicious.

A: As you ought to be! Extraordinary claims demand extraordinary evidence, and for exactly this reason the idea of time dilation has been tested by experiment many times. Remember that the effect is minute for clocks moving even as fast as the space shuttle, so the tests fall into two categories: tests by extremely accurate (man-made) clocks traveling at speeds far below the speed of light, and tests by natural clocks traveling at speeds near the speed of light.

In the first category are tests made using atomic clocks carried by commercial airliners (Hafele and Keating, 1972), carried by a rocket (Vessot and collaborators, 1980), and carried by a Navy P3C Orion antisubmarine patrol airplane (Alley, 1983). All of these tests have verified time dilation both qualitatively and quantitatively.

Even more tests in the second category have been performed. Here's one of them: A particle called a muon is a kind of a clock (in the same sense that your fingernail is a kind of a clock) because once a muon is created, it exists for a lifetime of 2.198 ± 0.002 microseconds and then decays into a electron.* To be precise, this is the muon's lifetime when it's at rest relative to the laboratory. In an experiment performed at CERN, muons were speeding past the laboratory at $V = 0.999\,418\,c$,

---

*The notation $\pm$ recognizes that every measurement involves some apparatus, and that no apparatus is perfect, so no measurement is exact. As far as current experiments can tell, the lifetime of a muon is about 2.198 microseconds, but it could be as long as 2.200 microseconds or as short as 2.196 microseconds.

so according to time dilation their lifetimes would be *longer*, namely

$$\frac{2.198 \pm 0.002 \text{ microsecond}}{\sqrt{1 - (0.999\,418)^2}} = 64.43 \pm 0.06 \text{ microsecond.}$$

And indeed, the measured lifetimes of the moving muons were 64.37 ± 0.03 microseconds. Those moving muon clocks were ticking slowly.

The highest accuracy test of time dilation yet performed involves clocks that are lithium ions, moving at speeds $0.03c$ and at $0.064c$, and the experimental results agreed with the prediction of the time dilation formula to $0.000\,000\,02\%$, which is smaller than the experimental error.

All of these tests were performed using clocks other than light clocks, demonstrating that time dilation is a property of time, not of any particular type of clock.

The book by Yuan Zhong Zhang lists 24 time dilation experiments, with clocks varying from human-made atomic clocks to the decay of muons to excited neon atoms to oscillating nuclei, moving at speeds from a few hundred miles/hour to $0.9994\,c$. So far, every experiment performed has verified time dilation to within the accuracy of the experimental apparatus. More experimental tests are in the planning stages: for example, the OPTIS satellite project is attempting to verify time dilation with one thousand times more accuracy than current tests.

Q: Does the moving clock tick slowly because it somehow got damaged in the jarring process of accelerating up to high speed?

A: No. Remember that the clock ticks off time correctly in its own frame. The acceleration process can be stretched out and made as gentle as needed to avoid damaging the clock.

Q: Suppose Veronica drives past Ivan at high speed. You say that her clocks tick slowly. Wouldn't she notice this herself?

A: As observed from Ivan's frame, Veronica's wristwatch ticks slowly, but also her heart beats slowly. Her lungs breath slowly, but we needn't fear for her life, because her cells respire slowly. Her thought processes proceed slowly as well. As I said, *all* the clocks tick slowly, and they all tick slowly by the same factor, which is the same as saying that time itself passes slowly.

Q: Suppose Veronica drives past Ivan at high speed. He says that her clocks are moving, and hence ticking slowly. But the principle of relativity tells us that Veronica's reference frame is just as good as Ivan's. She says that his clocks are moving, and hence ticking slowly. So he says that her clocks tick slowly, and she says that his clocks tick slowly. They can't both be right!

A: If time dilation were the only effect of relativity, this would indeed be a logical contradiction. This is our clue that there's more going on. Do you remember that at the beginning of this chapter we said that either time was screwy, or distance was screwy, or both? We've already discovered that time is screwy, and this apparent contradiction means that we'll have to go on and discover that distance is screwy too!

Q: The heroes of *Star Wars* often travel near the speed of light. Why do they never experience time dilation?

A: The *Star Wars* universe is only fictional. It's not as stunningly wild and exciting as our own, real, universe is.

PROBLEMS

4.1 Finding the square root factor. In the course of doing time dilation problems we will need to calculate the square root factor $\sqrt{1 - (V/c)^2}$ over and over again. To save labor, I have tabulated this factor for some velocities in Appendix D.

    a. Verify the left-hand half of this table.

    b. If $V = 20\,000$ miles per second, then what is $\sqrt{1 - (V/c)^2}$?

4.2 Time dilation examples. Verify the results for the four time dilation examples (with results of 52, 48, 36, and 8.5 minutes) given on page 34.

4.3 Time dilation for a biological clock. I trim my moustache every 6 weeks. If I were in a rocket ship flying past you at a speed of $V = \frac{4}{5}c$, then how much time, in your frame, would elapse between my moustache trimmings?

4.4 Muon lifetime. The lifetime of a stationary muon is 2.2 microseconds. A beam of muons is produced traveling at 83% of the speed of light. How far will the muons travel before they decay (a) if commonsense ideas of time were correct? (b) in the real world of relativity? (Measuring this distance is a direct test of time dilation, frequently performed in undergraduate physics teaching laboratories.)

REFERENCES

You can access the Boulder, Colorado, fountain atomic clock through www.time.gov. The September 2002 issue of *Scientific American* is devoted to articles about time and timekeeping.

The chapter epigraph is a translation of the message appearing on the French tapestry "Time" (c. 1505) (Cleveland Museum of Art, accession no. 1960.178). Einstein's remark on common sense (p. 35 above) appears in

E. T. Bell, *Mathematics: Queen and Servant of Science* (New York: McGraw-Hill, 1951), p. 42,

while Thoreau's remark on prejudice appears in

Henry David Thoreau, *Walden; or, Life in the Woods* (Boston: Ticknor and Fields, 1854).

The concept of time dilation, in its modern form, was introduced by

Albert Einstein, "Zur Elektrodynamik bewegter Körper," *Annalen der Physik*, 17 (1905) 891–921, trans. W. Perrett and G. B. Jeffery as "On the electrodynamics of moving bodies" and published in *The Principle of Relativity* (New York: Dover, 1952), pp. 35–65.

The light-clock style of derivation employed in this book was invented by

Gilbert N. Lewis and Richard C. Tolman, "Principle of relativity, and non-Newtonian mechanics," *Proceedings of the American Academy of Arts and Sciences*, 44 (1909) 711–24.

Tests of time dilation using man-made (atomic) clocks moving at relatively low speeds include

J. C. Hafele and R. E. Keating, "Around-the-world atomic clocks: Observed relativistic time gains," *Science*, 177 (1972) 168–70.
R. F. C. Vessot, M.W. Levine, E.M. Mattison, et al., "Tests of relativistic gravitation with a spaceffborne hydrogen maser," *Physical Review Letters*, 45 (1980) 2081–84.
Carroll O. Alley, "Proper time experiments in gravitational fields with atomic clocks, aircraft, and laser light pulses," pp. 363–427 in *Quantum Optics, Experimental Gravity, and Measurement Theory*, ed. Pierre Meystre and Marlan O. Scully (New York: Plenum Press, 1983).

All three of these tests compare two clocks that differ not only in speed but also in height, so they test not only the relativistic time dilation described here but also the gravitational time dilation described in chapter 17, "General Relativity."
    The CERN muon decay test is described in

J. Bailey, K. Borer, F. Combley, et al., "Measurements of relativistic time dilatation for positive and negative muons in a circular orbit," *Nature*, 268 (28 July 1977) 301–5.

The $^{7}Li^{+}$ ion test is described in

G. Saathoff, S. Karpuk, U. Eisenbarth, et al., "Improved test of time dilation in special relativity," *Physical Review Letters*, 91 (7 November 2003) 190403/1–4.
Sascha Reinhardt, Guido Saathoff, Henrik Buhr, et al., "Test of relativistic time dilation with fast optical atomic clocks at different velocities," *Nature Physics*, 3 (1 December 2007) 861–64.

The summary of 24 time dilation experiments is in

Y. Z. Zhang, *Special Relativity and Its Experimental Foundations* (Singapore: World Scientific, 1997).

The muon decay experiment mentioned in problem 4.4 is frequently performed in undergraduate physics teaching laboratories. See, for example,

Thomas Coan, Tiankuan Liu, and Jingbo Ye, "A compact apparatus for muon lifetime measurement and time dilation demonstration in the undergraduate laboratory," *American Journal of Physics*, 74 (2006) 161–64.

On page 31 I presented an argument showing that any kind of moving clock must tick slowly—whether it be a light clock, an electrical clock, a pendulum clock, a heart beat, a lengthening moustache, or a face slowly accumulating wrinkles as years pass by. Even so, it's interesting to delve into the exact mechanism through which each kind of moving clock ticks slowly. In some cases the mechanism has yet to be discovered. But the paper

Oleg D. Jefimenko, "Direct calculation of time dilation," *American Journal of Physics*, 64 (1996) 812–14.

works out the mechanism through which some moving electrical clocks tick slowly. If you remember that the heart maintains its steady rhythm through electrical circuitry (using the sinoatrial node and the atrioventricular node), then it might seem more plausible that moving hearts beat slowly.

CHAPTER FIVE

The Great Race

> Already I feel my brain reeling, and like a cloud suddenly cleft by
> lightning, it is troubled.
>
> —GALILEO, *TWO NEW SCIENCES* (1638)

A rocket sled takes a 300-foot straightaway race at constant speed $V = \frac{3}{5}c$.* With this speed, $\sqrt{1 - (V/c)^2} = \frac{4}{5}$. (The sled begins considerably behind the start line, accelerates to full speed, and then passes the start line already moving at speed $V = \frac{3}{5}c$. Similarly, it doesn't begin braking until after it passes the finish line. Therefore, we needn't worry about the acceleration and can concentrate on the uniform motion during the 300-foot straightaway race.) In the Earth's frame, this race requires a time of

$$\text{time elapsed} = \frac{\text{distance traveled}}{\text{speed}} = \frac{300 \text{ ft}}{\frac{3}{5}c} = \frac{300 \text{ ft}}{\frac{3}{5}(1 \text{ ft/nan})} = 500 \text{ nan.}$$

And indeed, clocks fixed at the start and finish line will tick off 500 nans during the race.

But the rocket sled's clock *doesn't* tick off 500 nans; that moving clock ticks slowly, so it ticks off a smaller time of $\frac{4}{5}(500 \text{ nan}) = 400 \text{ nan.}$ (This is *not* a defect of the sled's clock. It ticks off only 400 nans because, in the sled's frame, only 400 nans have elapsed.)

This figure summarizes the race as observed from the Earth's frame.

---

*Why a rocket sled, on runners, rather than a rocket car, on wheels? I'll answer this question in chapter 12, "Rigidity."

The great race from the Earth's frame

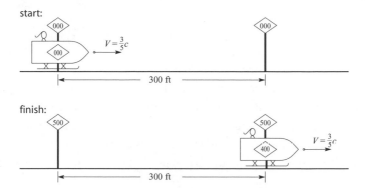

Now let's ask what happens in the race as observed from the sled's frame. Our first thought would be that the only difference is that, in the sled's frame, the sled is stationary and the start and finish lines move:

The great race from the sled's frame (first try)

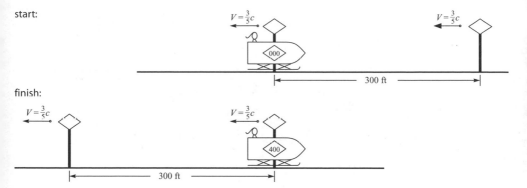

But there's something very wrong here. As observed from the Earth's frame, the sled approaches the finish line, 300 feet away, at speed $\frac{3}{5}c$. It gets there after 500 nans have elapsed. As observed from the sled's frame, the finish line approaches the sled, shown as 300 feet away, at speed $\frac{3}{5}c$. It gets there after 400 nans have elapsed. That's crazy! The finish line, moving at $\frac{3}{5}c$, can't cover 300 feet in 400 nans! In that amount of time the finish line can travel only

distance traveled = speed × time elapsed
$$= \tfrac{3}{5}c \times 400 \text{ nan} = \tfrac{3}{5}(1 \text{ ft/nan}) \times 400 \text{ nan} = 240 \text{ ft.}$$

We seem to have hit an insurmountable roadblock.

I admit without shame that if it had been me, rather than Einstein, uncovering relativity in 1905, this problem would have stumped me: I would have given up, discouraged, and gone out hiking. Einstein was not only smarter than I am (that goes without saying), but more importantly he was braver than I am. Instead of giving up, he took a risk and tried to resolve the issue.* Before I tell you Einstein's solution, can you take a shot at it yourself?

Einstein's solution is as simple and as necessary as it is outrageous. It is that in the sled's frame, the racetrack is only 240 feet long.

In the racetrack's own frame, it is 300 feet long, but in the frame in which the racetrack moves, it is shorter, as shown in this figure.

The great race from the sled's frame

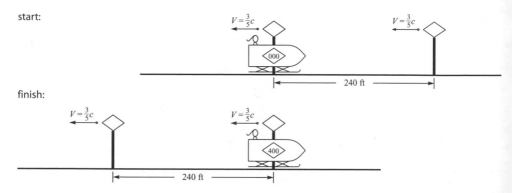

We executed this argument for a racetrack, but the racetrack was just scaffolding to propel the argument. If the race were run in a warehouse, the argument would

---

*Einstein didn't follow exactly the same reasoning that I'm presenting here, but the path he did take raised the same sort of issues and required the same sort of intelligence and ingenuity and intellectual courage.

show that a moving warehouse is short. If the race were run along the side of a truck, the argument would show that a moving truck is short. In general,

A moving rod is short.

This phenomenon is called *length contraction*.

At the Finish Line of the Great Race

The rocket sled passes the finish line of the great race, the checkered flag waves, the sled's clock reads 400 nans, and the finish-line clock reads 500 nans. We understand this thoroughly from the Earth's point of view.

But what about from the sled's point of view? In the sled's frame, the race lasts only 400 nans. Furthermore, the moving finish-line clock has been ticking slowly, so during the race it hasn't ticked off 400 nans, but a smaller time, $\frac{4}{5}(400 \text{ nan}) =$ 320 nan. How can a clock tick off 320 nans yet read 500 nans?

Having seen how Einstein resolved the "length of the moving racetrack" problem, can you take a guess at how to resolve this one?

There is only one way. A clock that ticks off 320 nans and reads 500 nans must have started off reading 180 nans. When the sled crossed the start line at the beginning of the race, and the start-line clock read 000 nans, the finish-line clock must have read, not 000 nans, but 180 nans. In the last two figures I showed the start and finish lines, but not their clock readings. Here I restore them:

The great race from the sled's frame

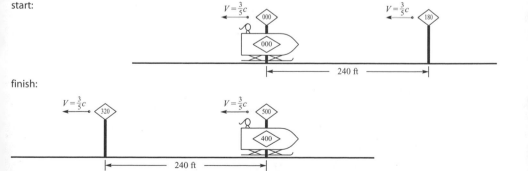

Yes. In the Earth's frame, the start-line clock and finish-line clock are both at rest, and they are synchronized. But in the sled's frame, these two clocks are moving, and they are not synchronized: the finish-line clock, the clock to the rear, is set ahead.

A moving pair of clocks isn't synchronized; the rear clock is set ahead.

This phenomenon is called *the relativity of synchronization* (or, sometimes, the relativity of simultaneity).

Q: Suppose Ivan has a pair of clocks, stationary and synchronized in his own frame. Veronica runs to the right past Ivan. In her frame, the pair of clocks moves to the left, so the right-hand clock is the rear clock and is set ahead. Veronica's sister Denise runs to the left past Ivan. What is the situation in Denise's frame?

A: In Denise's frame, the pair of clocks moves to the right, so the left-hand clock is the rear clock and it is set ahead.

Ivan's frame

Veronica's frame

Denise's frame

Q: So you claim that in Ivan's frame, the two clocks are synchronized, while in Veronica's frame the right-hand clock is set ahead and in Denise's frame the right-hand clock is set behind. How can the right-hand clock be both set ahead and set behind at the same time?

A: That phrase "at the same time" is causing trouble. In Ivan's frame, the two clocks strike noon simultaneously. In Veronica's frame, the right-hand clock strikes noon first, and some time later the left-hand clock strikes noon. In Denise's frame the two clocks strike noon in the opposite sequence. So in two of the three frames, these events don't happen "at the same time."

We've already seen that to specify a speed correctly we must specify that speed with respect to some particular reference frame. Then we found that to specify a time interval or a length we had to specify it with respect to some particular reference frame. Now we find that to specify a sequence we have to specify that sequence with respect to some particular reference frame.

The English language was invented by people who were not familiar with relativity, so it often works against our understanding. There's a word for "before" and a word for "after," but there's no word for "maybe before or maybe after or maybe at the same time, depending on reference frame." Our commonsense notions of space and time have been built into our language, but our commonsense notions of space and time are wrong. We have to be on our guard because our language encourages us to say (and to think!) false things.

Q: Time dilation and length contraction are weird, but I can live with their weirdness. I can imagine (if I stretch my mind) how a clock might gradually slow down as its speed increases, or how a rod might gradually shrink as its speed increases. I'm not familiar with this, but I can see how I might become familiar if I just gave the concepts time to sink in.

But this clock synchronization business is just too weird. I can't see how to get from here to there. Suppose a stationary pair of clocks starts out synchronized, and then the speed of each clock is gradually increased. As the clocks move faster and faster, they tick more and more slowly. But each clock travels at the *same* speed, and therefore each clock accumulates the *same* slowness. How can the two clocks fall out of synchronization?

A: You're in good company with this question: Einstein asked it too. I'm going to put off your question until we're more familiar with uniform motion in a straight line. Only then, in chapter 18, "A Pair of Clocks Starts Moving," will we move on to answer questions like yours about acceleration.

PROBLEMS

5.1 ✦ Another race. Veronica takes a 195-foot straightaway race at the constant speed of $V = \frac{5}{13}c$. There is a clock at the start line, and a clock at the finish line, and Veronica wears a wristwatch. As Veronica passes the start line, her wristwatch and the start-line clock both read zero. (The start- and finish-line clocks are synchronized in the Earth's frame.)

a. In the Earth's frame, how much time elapses during the race?
b. What does Veronica's watch read when she crosses the finish line?
c. In Veronica's frame, how long is the racetrack?
d. In Veronica's frame, how much time does the finish-line clock tick off during her race?
e. In Veronica's frame, what is the reading on the finish-line clock at the start of the race?

CHAPTER SIX

Length Contraction

T his chapter and the next are nearly the same as chapter 5, "The Great Race," except that here I use symbols instead of numbers, so I produce formulas for length contraction and the relativity of synchronization. This way, instead of just saying that "a moving rod is short," I'm able to say exactly how short it is.

A rocket sled takes a straightaway race of length L_0 at constant speed V. (The sled begins considerably behind the start line, accelerates to full speed, and then passes the start line already moving at speed V. Similarly, it doesn't begin braking until after it passes the finish line.) In the Earth's frame, this race requires a time of

$$\text{time elapsed} = \frac{\text{distance traveled}}{\text{speed}} = \frac{L_0}{V}.$$

And indeed, clocks fixed at the start and finish line will tick off a time of L_0/V during the race.

But the rocket sled's clock *doesn't* tick off a time of L_0/V: that moving clock ticks slowly, so it ticks off a smaller time of $\sqrt{1 - (V/c)^2}L_0/V$.

Compare what happens in the sled's frame with what happens in the Earth's frame:

From the Earth's point of view, the sled approaches the finish line at speed V. The distance it travels is

$$\text{distance traveled} = \text{speed} \times \text{time elapsed}$$
$$= V \times (L_0/V)$$
$$= L_0.$$

From the sled's point of view, the finish line approaches the sled at speed V. The distance it travels is

distance traveled = speed × time elapsed
$$= V \times (\sqrt{1 - (V/c)^2}L_0/V)$$
$$= \sqrt{1 - (V/c)^2}L_0.$$

That is, the finish line moves a smaller distance in the sled's frame than the sled moves in the Earth's frame. The racetrack itself must be shorter in the sled's frame.

A rod that has length L_0 in its own frame has a shorter length $\sqrt{1 - (V/c)^2}L_0$ in a frame in which it moves at speed V.

This phenomenon is called *length contraction* (or, sometimes, Lorentz contraction or Lorentz-Fitzgerald contraction). The length L_0 is called the rod's "rest length."

I need to warn you about language. I have said that a rod with length L_0 as observed from its own frame has a shorter length, L, as observed from another frame. Often this result is stated as "A rod with length L_0 as observed from its own frame appears to have a shorter length, L, as observed from another frame." This statement is true: the rod appears to have shorter length, L, because it *does* have shorter length, L. Using the term "appears" gives the false impression that, when the rod is observed from a frame in which it moves, the rod *really is* of length L_0 and only *appears to be* of length L. No. As observed from a frame in which it moves, the rod *really does* have the shorter length L. As observed from its own frame, the rod *really does* have the longer length L_0. Statements like "A rod has length L_0 in its own frame, but it appears to have a shorter length, L, in another frame" foster the same misimpression as statements like "A car has speed 0 miles/hour in its own frame, but it appears to have speed 60 miles/hour in the Earth's frame" or "New York City is 500 miles from Cleveland, but it appears to be 3000 miles from San Francisco."

Q: But the moving rod is not really shorter, right? It just seems to be shorter because it takes some time for the light to reach your eye.
A: The moving rod really is shorter. The "old light" effect that you bring up is real and does happen,* but it's not what we're talking about here. For example, suppose that Cynthia, standing on the shoulder of a highway, watches a truck approach. Light from the nose of the truck takes some time to reach Cynthia's eye, but light from the tail of the truck takes even more time to reach her eye, because

---

*We discussed it in problem 3.1, "Looking at a clock," on page 22.

it has further to travel. Thus, Cynthia sees the nose of the truck as it was some time ago, but the tail of the truck as it was even more time ago.

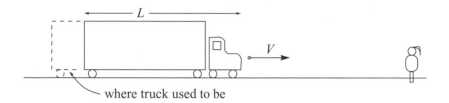

where truck used to be

For this reason, Cynthia sees the truck to be *longer* than it really is. This is not length contraction: Length contraction is about how long the rod *is*, not about how long it *appears to be*.

Q: Does the moving rod shorten because it's compressed by the air hitting it?

A: No. Remember that the rod still has its ordinary rest length in the frame that moves along with the rod.

Q: Does the moving (short) rod get fatter as atoms rearrange themselves within the rod?

A: No, the rod maintains its same width in both reference frames. The atoms don't rearrange themselves: instead, the moving atoms are themselves foreshortened (so that they are shaped sort of like door knobs rather than sort of like tennis balls), and the distance between atoms (in the direction of motion) is decreased. Any rod (like any street scene, page 5) can be observed either from a frame in which it's at rest or from a frame in which it moves. The density (atoms per volume) of that single rod depends on the reference frame: it's greater in the frame in which the rod moves (see fig. 6.1).

Q: If a moving rocket sled is short, and a moving clock is short, then why don't you show that in your sketches?

A: If you compare the Earth-frame sketch on page 41 with the sled-frame sketch on page 43, you'll see that the moving sled in the Earth's frame is 4/5 the length of the resting sled in the sled's own frame. The moving clock and the moving face are shrunken by the same factor.

Q: After you introduced time dilation, you backed up your argument by appealing to dozens of experimental tests of time dilation. How about length contraction?

A: No one has ever made a direct experimental test of length contraction by itself. Here's why: The space shuttle is 122 feet long. Suppose it passes us at its orbital speed of 5 miles/second. According to the length contraction formula,

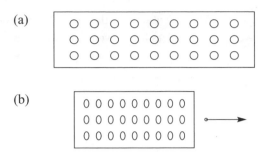

Fig. 6.1. (a) A rod from its own reference frame (i.e., from the frame in which the rod is at rest). Representative atoms are shown as circles. (b) The same rod from a reference frame in which it moves at $V = \frac{4}{5}c$. The atoms are short in the direction of their motion.

it is contracted by 2×10^{-8} meter—about the width of 100 atoms. This same contraction would result from reducing the shuttle's temperature by 0.0001 degree Celsius. It is difficult to measure any length to this level of accuracy, much less the length of something that's speeding by at 5 miles/second!

While there has never been an experimental test of length contraction by itself, we will find, on pages 92 and 101, that there have been many (literally billions) of experimental tests of the package of time dilation, length contraction, and the relativity of synchronization.

Q: If there's been no direct test of length contraction, then why should I think it happens?

A: There are many scientific conclusions for which the evidence is indirect rather than direct. No one has ever seen an atom, or the surface of Venus, or the Earth's iron core, or the muscles of a dinosaur. No one has ever seen a radio wave. Before 12 April 1961, when Yuri Gagarin became the first space pilot, no one had ever seen the Earth as a sphere. If you're like me, then no one has ever seen your liver, or your kidneys, or your heart. The evidence for these phenomena is no less impressive for being indirect.

Q: I previously raised the objection (page 36) that Ivan says Veronica's clocks tick slowly, while Veronica says Ivan's clocks tick slowly. You said this apparent contradiction meant only that we had more work to do. We've done more work: Now we know that Ivan says Veronica's rods are short, while Veronica says Ivan's rods are short. Instead of resolving the paradox you've doubled it!

A: Proof that we have still more work to do . . .

CHAPTER SEVEN

Clock Synchronization

The rocket sled passes the finish line of the great race, the checkered flag waves, the sled's clock reads time T_S, and the finish line clock reads a longer time, T_F ($T_S = \sqrt{1 - (V/c)^2}\,T_F$). We understand this thoroughly from the Earth's point of view.

But what about from the sled's point of view? In the sled's frame, the race lasts only for time T_S. Furthermore, the moving finish-line clock has been ticking slowly, so during the race it hasn't ticked off time T_S, but a smaller time,

$$\sqrt{1 - (V/c)^2}\,T_S = \sqrt{1 - (V/c)^2}(\sqrt{1 - (V/c)^2}\,T_F) = (1 - (V/c)^2)T_F.$$

How can a clock tick off a time of $(1 - (V/c)^2)T_F$ yet read the larger time T_F?

There is only one way. A clock that ticks off time $(1 - (V/c)^2)T_F$ and reads time T_F must have started off reading

$$T_F - (1 - (V/c)^2)T_F = \frac{V^2}{c^2}T_F = \frac{VL_0}{c^2}.$$

Yes. In the Earth's frame, the start-line clock and finish-line clock are both at rest, and they are synchronized. But in the sled's frame, these two clocks are moving, and they are not synchronized: the finish-line clock, the clock to the rear, is set ahead.

A pair of clocks, synchronized in the pair's own frame, is not synchronized in a frame in which the pair moves at speed V. In that frame the rear clock is set ahead by L_0V/c^2.

This phenomenon is called *the relativity of synchronization* (or, sometimes, the relativity of simultaneity).

Q: I'm confused. The time dilation formula relates T_0 and T. But in the situation you just described, the times were T_S and T_F. Why did you introduce these new symbols? Is T_S equal to T_0 or to T?

A: That depends. T_S is the time ticked off by the sled's clock—that is, the time elapsed in the sled's frame. T_F is the time ticked off by the finish-line clock—that is, the time elapsed in the Earth's frame. In the sled's frame, T_0 means T_S and T means T_F, but in the Earth's frame, T_0 means T_F and T means T_S.

There's a strong desire to solve physics problems by "finding the right equation and plugging in." I'm pleased to say that this technique doesn't work: you have to think.

Q: Suppose a pair of clocks is separated by 5 feet in the pair's own frame, but (due to length contraction) by a shorter distance, 3 feet, in the frame in which the clocks move at speed V. In the formula L_0V/c^2, does L_0 mean 5 feet or 3 feet?

A: It means 5 feet, the distance between the two clocks in the pair's own frame.

What Does "Rear" Mean?

The word *rear*, as used in relativity, is subtly different from the everyday meaning of the word. Consider this situation: As observed from the Earth's frame, a car moves right at 20 miles/hour and a motorcycle moves right at 50 miles/hour. In the Earth's frame, the car moves right: its headlamps are to the front and its tail lamps are to the rear. But in the motorcycle's frame, the car moves left at 30 miles/hour: its tail lamps are to the front and its headlamps are to the rear. And in the car's own frame, the car is stationary; neither end of the car is rear. Because of this subtle difference, some find it easier to think "the *trailing* clock is set ahead" rather than "the *rear* clock is set ahead."

In the great race from the Earth's frame (page 41) the start-line/finish-line pair of clocks is not moving, so neither clock is to the rear. But from the sled's frame (page 43) this pair of clocks moves left, so the right-hand clock is to the rear. We will later encounter a pair of clocks mounted in a railroad coach (page 103). From the coach's frame this pair of clocks is not moving, so neither clock is to the rear. But from the Earth's frame the coach moves right, so the left-hand clock is to the rear.

Earth's frame

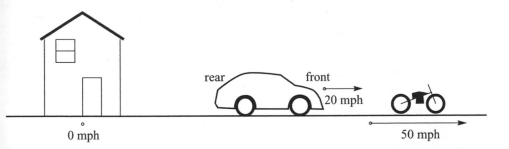

Car's frame

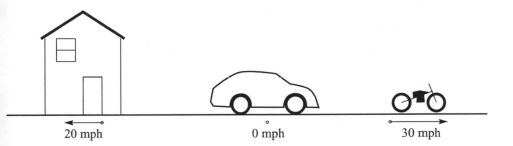

Motorcycle's frame

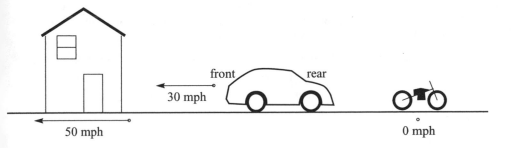

PROBLEMS

7.1 ✦ The meaning of "rear." A sedan moves relative to the road at 20 miles/hour. A sports car is catching up with it by driving at 50 miles/hour (relative to the road). Is the sedan's trunk to the rear of its hood

    a. in the Earth's frame?

    b. in the sports car's frame?

    c. in the sedan's frame?

7.2 ✦ Engine trouble. In the frame of a moving car, the driver coughs, and at the same moment (by coincidence) a puff of smoke emerges from the tailpipe. In the Earth's frame which is true?

    a. These two events are again simultaneous.

    b. The puff of smoke emerges before the driver coughs.

    c. The driver coughs before the puff of smoke emerges.

    d. The driver never coughs.

Explain your answer in one sentence.

7.3 Three clocks. Ivan sets three clocks onto the ground and makes sure they're synchronized. Clock B is 30 feet to the right of clock A, and clock C is 60 feet to the right of clock B. Veronica runs to the right at speed $V = \frac{1}{5}c$ relative to Ivan. In Veronica's frame, when clock A reads 100 nans, what are the readings on clocks B and C?

7.4 ✦ Two clocks. Ivan sets two clocks onto the ground and makes sure they're synchronized. Clock B is 30 feet to the left of clock A. Veronica runs to the right at speed $V = \frac{1}{5}c$ relative to Ivan. In Veronica's frame, when clock A reads 100 nans, what is the reading on clock B?

7.5 ✦ Boxcar with a bomb. A railroad boxcar has a clock mounted on its front wall and a clock mounted on its rear wall, and a package at its very center. Each clock is attached to a flashlight aimed at the package, and each flashlight is programmed to flash momentarily when its clock strikes noon. Within the package is a bomb set to explode if it gets a light signal from one flashlight or from the other, but *not* if it gets a signal from both. The boxcar travels down a track at high speed.

Susan analyzes this situation from the boxcar's frame: "The two flashlights flash simultaneously, the two light signals travel the same distance, and the two light signals reach the package simultaneously. Therefore, the bomb *does not* explode."

Roth analyzes this situation from the Earth's frame: "The two flashlights flash simultaneously, but the bomb is moving away from the rear (left-hand) light signal and toward the front (right-hand) light signal. The front light signal thus has less distance to travel, so it arrives at the package first. Therefore the bomb *does* explode."

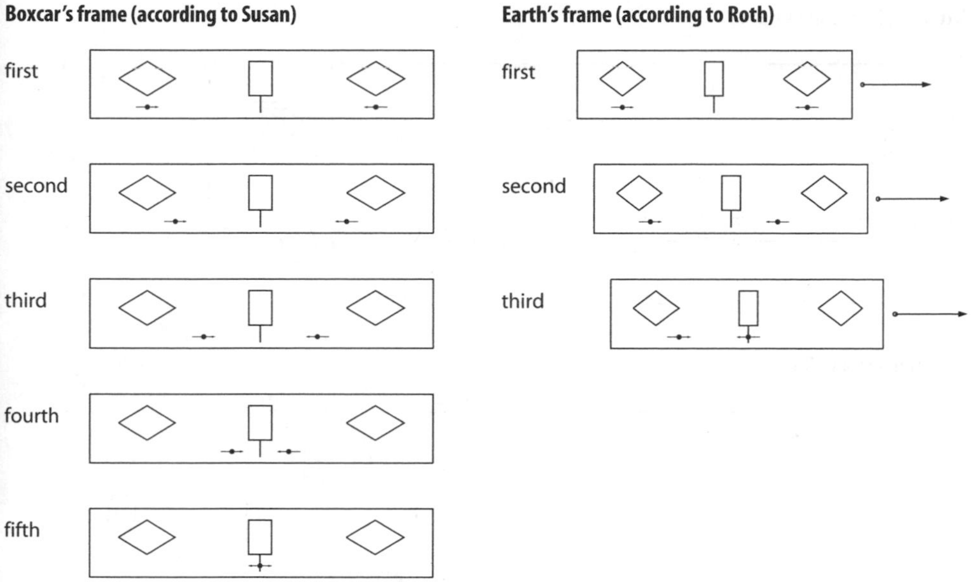

Boxcar's frame (according to Susan)   **Earth's frame (according to Roth)**

first   first

second   second

third   third

fourth

fifth

Clearly, these two analyses cannot both be correct. Find the flaw (a hidden assumption) in one analysis, and state definitively whether the bomb explodes or not.

7.6 Train in a tunnel. A train with skylights has the same rest length as a tunnel. The train goes through the tunnel at high speed. In the train's frame, the tunnel is contracted to be shorter than the train. As the train passes through the tunnel, two passengers, Wendy at the front and John at the rear, glance up at the same moment, and both of them see blue sky. In the tunnel's frame, the train is contracted to be shorter than the tunnel. How can the two passengers look up and both see blue sky?

(If you find this problem too murky and fluid, you may render it concrete and tangible by considering a train and tunnel, each of rest length 500 feet, with relative speed $V = \frac{3}{5}c$. In the train's frame, Wendy and John glance upward when their wristwatches read 0 nan. What happens in the tunnel's frame? Your resolution will require one sketch in the train's frame, and two in the tunnel's frame.)

7.7 Veronica in a semitrailer. Veronica drives past Ivan at $V = \frac{3}{5}c$ in a semitrailer carrying three clocks.

Situation in Veronica's frame

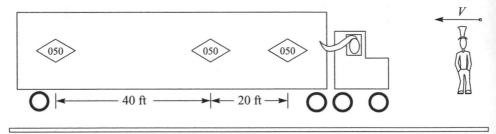

Situation in Ivan's frame

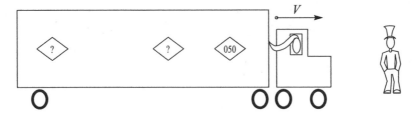

a. In Ivan's frame, what does the middle clock read?

b. In Ivan's frame, what does the left-hand (rear) clock read?

c. In Ivan's frame, how far apart are the front and rear clocks?

Resumé

Q: You started off by claiming that the speed of light was the same in all inertial reference frames. I was suspicious but I followed along. Pretty soon this claim got you into trouble, but you avoided it by inventing time dilation, which made me even more suspicious. But that wasn't the end of it. Time dilation got you into hot water, which you could only get out of by inventing length contraction and the relativity of synchronization. How many more times are you going to pull rabbits out of hats? How many more times will you expect me to grasp the implausible? When is all of this going to end?

A: It ends right now. With these three principles—time dilation, length contraction, and the relativity of synchronization—we have uncovered the core of special relativity. There will be new applications and new consequences as we explore these three ideas and become familiar with them, but no new core ideas. You have reached a milestone in your understanding of nature, and a pinnacle of human achievement.

Q: In that case, I'm disappointed. I was expecting something grander from science, some window to the meaning of life, something that would address great questions of Truth, Goodness, and Beauty.

A: These are indeed great and important questions, but they fall outside the purview of science. Science is the investigation of nature through reason, building from observation and experiment. By its very character, it can't answer questions that involve "should" or "ought." These questions might be very grand ("What is the character of justice?") or very important to you personally ("Whom ought I marry?"), or they might be utterly trivial ("What should be my favorite color?" "What should I eat for dinner tonight?"), but in any case they're not scientific questions. There are some who think that science is going to creep into every part of every person's life, eliminating passion and fun and mystery and spontaneity and rendering religion, art, and philosophy impotent. It's not going to happen. Science is very good at what it does, but it doesn't and can't answer questions of morality, humaneness, and love.*

Q: But what about the war between science and religion?

A: Most of what you hear about the "war between science and religion" is nonsense. Science and religion are about different things: science is about what is, religion is about what ought to be. When these subjects reach outside of their bounds—when theologians try to make the universe conform to the Bible, or when scientists try to dictate morality—then there's trouble. But if both disciplines keep their limitations in mind, then a "war between science and religion" makes no more sense than does a "war between pork chops and the color blue."

Einstein said it better than I can: "Science without religion is lame, religion without science is blind." (It would take us too far afield to uncover exactly what Einstein meant by religion in this sentence, but he definitely did *not* mean a set of rituals, myths, and dogmas to be trotted out one day a week and otherwise ignored.)

---

*This is not to say that science is useless. For example, science can't tell you whether your favorite color should be black or white, but it can tell you that on a sunny day a black coat will be warmer than an equivalent white coat. Science can't tell you whether your favorite food should be beef or chicken, but it can tell you how beef and chicken will affect your chances of a heart attack. Science can't tell you that one objective is better than another, but it can help you reach your objective once you've selected it.

Is Relativity Complicated?

The theory of relativity has a reputation for being complex and impenetrably difficult. "Sure I'll learn how to understand my telephone bill," a friend of mine once
announced sarcastically, "just as soon as I've mastered Einstein's theory of relativity."
As soon as I got home I looked at my telephone bills. My land line bill contained
an inscrutable "Access recovery charge" (How does one recover access?), an "Interstate access charge" (despite the fact that I had made no interstate calls), and a "Svc
Provider Number Portability Fee" (What is Svc? What is a portable number, and why
should I pay for one?). My cell phone bill listed features like ANYTIME ROLLOVER
MINS, UNL Night & Wknd Min, UNLTD EXP M2M MINS, and NT/WKNDS SHRD (cap-
italization unaltered from the originals). It indeed required esoteric knowledge to
decipher these bills.

In contrast, here is relativity:

A moving clock ticks slowly.
A moving rod is short.
A moving pair of clocks isn't synchronized.

Three short and straightforward assertions.

Counterintuitive? Yes.
Profound? Yes.
Complicated? No.

Compared with the complex and arcane rules of telephone accounting, relativity is
simplicity incarnate.

REFERENCES

In the text I said there was trouble when either religion or science overstepped its
bounds. A case of religion overstepping is described in

Dava Sobel, *Galileo's Daughter* (New York: Walker and Co., 1999),

and a case of science doing the same appears in

Stephen Jay Gould, *The Mismeasure of Man* (New York: Norton, 1981).

Both oversteppings had heinous consequences for religion, for science, and for the
people touched.

Einstein's comment on science and religion was made in an address delivered on 11 September 1946 and published in

Albert Einstein, "Science and Religion," p. 211 in *Science, Philosophy, and Religion: A Symposium* (New York, 1941).

PART III / Exploring Relativity

CHAPTER EIGHT

The Case of the Hungry Traveler

There were [towering] clouds to the south, drifting over the Hopi
Reservation. The Hopis had held a rain dance Sunday, calling on the
clouds—their ancestors—to restore the water blessing to the land.
Perhaps the [clouds] had listened to their Hopi children. Perhaps not. It
was not a Navajo concept, this idea of adjusting nature to human needs.
The Navajo adjusted himself to remain in harmony with the universe.
When nature withheld the rain, the Navajo sought the pattern of this
phenomenon—as he sought the pattern of all things—to find its beauty
and live in harmony with it.

—TONY HILLERMAN

We have uncovered the three big ideas of special relativity: time dilation, length
contraction, and the relativity of synchronization. But simply knowing what
these ideas are doesn't mean we know how to work with them or use them, any more
than simply knowing a person's name means you have become friends. Our goal now
is to become familiar with these ideas, to apply them in various circumstances, to
see where they lead and what they have to teach us, and, in short, to become friends
with them.

Residents of Rhode Island grow up familiar with the idea of traveling to a different
state to take advantage of lower grocery prices. When a long-time resident of Rhode
Island visits Texas, he is invariably surprised by the immensity of that state. Certainly,
from elementary school onward, he has been able to recite, "Texas is the second-
largest state in the Union," but that recitation does not prepare him to drive at
high speed for two full days through lush swamp woodlands, fertile ranches, rolling
savannas, blooming meadows, parched mountains, and High Plains deserts before
reaching the far side of the state. To recite "A moving clock ticks slowly" is no great
achievement, but to understand the implications and the power and the beauty of
this simple statement is a great achievement.

The next nine chapters work to build this understanding and familiarity. We begin,
in this chapter, by showcasing the three ideas in a concrete situation, written as a
series of six solved problems.

Problem 1: When her wall clock reads noon, Rosemary leaves her home and travels to her favorite deli at a speed of $V = \frac{3}{5}c$. Her wristwatch records 120 nans for her trip. What time does the deli's clock read when she arrives? All three clocks involved keep excellent time. (Tip: If $V = \frac{3}{5}c$, then $\sqrt{1 - (V/c)^2} = \frac{4}{5}$.)

Solution 1(A): Rosemary's moving clock ticks slowly, so the home and deli clocks record a longer time: $\frac{5}{4}(120 \text{ nan}) = 150$ nan. In the Earth's frame, this is the time elapsed during the journey.

Solution 1(B): Alternatively, use the formula

$$T = \frac{T_0}{\sqrt{1 - (V/c)^2}}$$

and realize that T_0 is the time ticked off by the single moving clock, Rosemary's clock, while T is the time obtained through the use of the two Earth-bound clocks.

Problem 2: In the Earth's frame, how far is the deli from Rosemary's home?

Solution 2: It is the distance Rosemary traveled, namely

distance traveled = speed × time elapsed
$$= \tfrac{3}{5}c \times 150 \text{ nan} = \tfrac{3}{5}(1 \text{ ft/nan}) \times 150 \text{ nan} = 90 \text{ ft.}$$

The situation in the Earth's frame is summarized in figure 8.1.

Problem 3: Now we seek to understand the journey as observed from Rosemary's frame. In Rosemary's frame, how far is the deli from Rosemary's home?

Solution 3(A): The distance in Earth's frame (the rest frame of the two buildings) is 90 feet. The distance in Rosemary's frame is a length-contracted $\frac{4}{5}(90 \text{ ft}) = 72$ ft.

Solution 3(B): In Rosemary's frame, Rosemary is stationary and the deli moves toward her at speed $V = \frac{3}{5}c$. The deli requires an elapsed time of 120 nans to reach her, so it travels a distance of

distance traveled = speed × time elapsed
$$= \tfrac{3}{5}c \times 120 \text{ nan} = \tfrac{3}{5}(1 \text{ ft/nan}) \times 120 \text{ nan} = 72 \text{ ft.}$$

Notice that questions like "How far is the deli from Rosemary's home?" and "How much time did the journey take?" are incomplete questions. One must ask instead, "How much time did the journey take in the Earth's frame?" or "How much time did the journey take in Rosemary's frame?" Because these are two distinct questions, it isn't surprising that there are two distinct answers. (Namely 150 nans and 120 nans, respectively.) In exactly this way, the question "How far

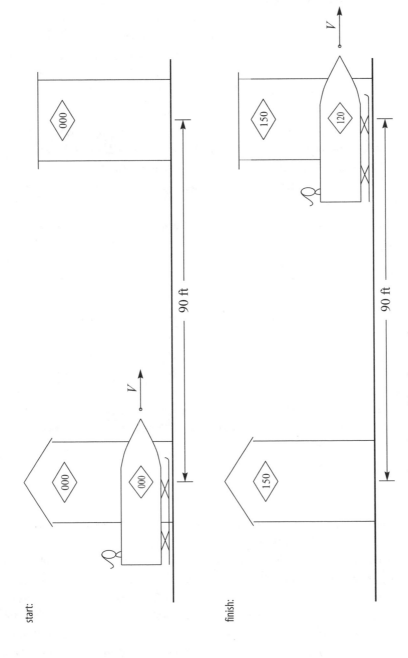

Fig. 8.1. Rosemary's trip to the deli, from the Earth's reference frame.

away is San Francisco?" is incomplete. One must ask instead, "How far away is San Francisco from Tokyo?"

Problem 4: You know that Rosemary leaves her home at noon and travels for 120 nans (as recorded by her wristwatch), and that when she finishes her journey the deli clock reads 150 nans after noon. The deli manager finds this quite natural: "Rosemary, while you were traveling, your wristwatch was ticking slowly," he explains. But Rosemary doesn't see it that way: "Listen, bub, *I've* been stationary [in my own frame]. *You're* the one who's been traveling. *Your* clock was ticking slowly all the while it was rushing toward me. In fact, while my wristwatch ticked off 120 nans, your traveling deli clock ticked off only $\frac{4}{5}$(120 nan) = 96 nan. Do you want to fight about it?" Defuse this tense situation by telling Rosemary the reading on the deli clock (in Rosemary's frame) at the instant that Rosemary's wall clock at home struck noon.

Solution 4: In Rosemary's frame the deli clock (the rear clock) is set ahead of the home clock by

$$\frac{L_0 V}{c^2} = \frac{(90 \text{ ft})(\frac{3}{5}c)}{c^2} = \frac{(90 \text{ ft})\frac{3}{5}}{c} = \frac{(90 \text{ ft})\frac{3}{5}}{1 \text{ ft/nan}} = 54 \text{ nan.}$$

So, in Rosemary's frame, the deli clock was set to 54 nans when the journey started, then it ticked off 96 nans, so of course it reads 54 nan + 96 nan = 150 nan when Rosemary arrives.

Problem 5: In Rosemary's frame, what does Rosemary's home clock read when Rosemary arrives at the deli?

Solution 5: During Rosemary's journey her home clock ticks off 96 nans, just as the deli clock does, so when she arrives at the deli her home clock reads 96 nans after noon.

The situation in Rosemary's frame is summarized in figure 8.2.

Q: Hold on! Which picture is right? When Rosemary arrives at the deli, does her home clock read 150 nans or 96 nans?

A: That depends. In the Earth's frame the home clock reads 150 nans when Rosemary arrives at the deli. In Rosemary's frame the home clock reads 96 nans when Rosemary arrives at the deli. So in the Earth's frame the two events "Rosemary arrives at deli" and "home clock reads 150 nans" are simultaneous. But in Rosemary's frame those two events aren't simultaneous: the rear event ("Rosemary arrives at deli") happens first, and then some time later the front event ("home

Rosemary's frame

start:

finish:

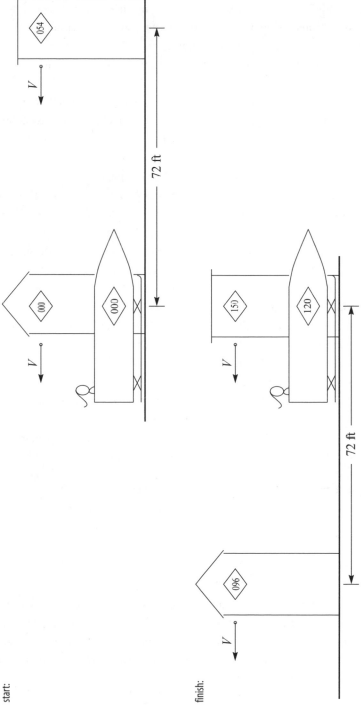

Fig. 8.2. Rosemary's trip to the deli, from Rosemary's reference frame.

clock reads 150 nans") occurs. The two figures depict the same journey, merely observed from two different reference frames. Both pictures are right.

Q: But that's contrary to common sense!

A: Of course. We don't commonly travel at $V = \frac{3}{5}c$, more than 100 000 miles/second.

Q: Well, maybe you're right, but suppose I ask a more concrete question. What if the deli manager looks at Rosemary's home clock, perhaps using binoculars. When Rosemary arrives at the deli, will he see the home clock reading 150 nans or 96 nans after noon?

A: Neither. It takes some time for the light from the home clock to reach the deli manager, so he will see the "old light" that left Rosemary's wall clock some time ago. I'm glad you raised this objection, because it makes a great problem:

Problem 6: Deli manager Aaron watches Rosemary's home clock through binoculars. What clock reading does he see when Rosemary arrives at the deli? Provide an analysis in both the Earth's frame and Rosemary's frame.

Solution 6(A) (analysis in the Earth's frame): The light travels 90 feet from Rosemary's home to Aaron's binoculars, and this requires 90 nans. Thus, Aaron sees the light that left the home clock 90 nans ago, when that clock read 150 nan − 90 nan = 60 nan.

Solution 6(B) (analysis in Rosemary's frame): This situation is more complicated because the light doesn't travel 72 feet. That's because Aaron is traveling left to meet the light emitted by the home clock, so the light travels less than 72 feet. Indeed, if the time required for the light to fly from Rosemary's home clock to Aaron's binoculars (in Rosemary's frame) is t_f, then

distance traveled by light + distance traveled by Aaron = 72 ft

$$ct_f \qquad + \qquad \tfrac{3}{5}ct_f \qquad = 72 \text{ ft.}$$

That is,

$$\tfrac{8}{5}ct_f = 72 \text{ ft} \quad \text{or} \quad t_f = \tfrac{5}{8}\frac{72 \text{ ft}}{1 \text{ ft/nan}} = 45 \text{ nan.}$$

Thus, the light that reaches Aaron at time 120 nans left Rosemary's home 45 nans ago. But while 45 nans elapse, the moving home clock ticks off less than 45 nans: in fact, it ticks off $\frac{4}{5}(45 \text{ nan}) = 36 \text{ nan}$. Hence, Aaron sees a clock reading 96 nan − 36 nan = 60 nan, which is the same result we found through our analysis in the Earth's frame.

The two different analyses disagree on issues like "How long does it take light to fly from Rosemary's home clock to Aaron's binoculars?" (It takes 90 nans in the Earth's frame and 45 nans in Rosemary's frame.) But they both agree on the observational consequence, "Aaron peers into his binoculars and sees a clock reading 60 nans," as they must!

Caution! In the everyday world these four phrases have the same meaning:

"the situation from Aaron's point of view"

"the situation in Aaron's frame"

"the situation as observed from Aaron's frame"

"the situation that Aaron sees"

But in the real world, the world of relativity, the last phrase has a very different meaning. What Aaron sees is different from what is happening, because it takes time for light to reach Aaron's eye. (In the context of relativity "see" means simply to look, while "observe" means to look and also take account of the light-travel delays, thus figuring out what is really happening right now.)

Q: What would happen if Rosemary's home clock emitted not light, which travels at speed c, but the newly invented "insta-rays," which (according to the inventor) travel from the home to the deli instantaneously?

A: In that case there would indeed be a contradiction: there would be no way to determine whether the home clock read 150 nans or 96 nans after noon. But according to special relativity, it makes no sense to say that insta-rays can send signals instantaneously: what's "instant" in one frame requires time to elapse in another frame. (In other words, two events simultaneous in one frame will not be simultaneous in another frame.)

Indeed, we will see in chapter 10, "Speed Limits," that no signal can travel faster than the speed of light! I advise you not to invest in the "insta-ray" corporation; according to special relativity, its product is a fraud.

Q: (This question is somewhat esoteric, and you might want to skip it on a first reading.) Given that it takes time for light to travel from its source to the eye (or to the binoculars or to the camera), how could pictures like figure 8.1 or 8.2 ever be made?

A: The easiest way would be to build a road parallel to the "home-to-deli road" but 12 000 feet away. Place a camera on this road directly in front of the home and have it snap a photo exactly 12 000 nans after Rosemary leaves her home to give the light time to travel from Rosemary's home to the camera. The result will be very much like the top sketch in figure 8.1. (Not exactly the same, because the deli is 12 000.3 feet from the camera, so light reaching the camera from the deli is 0.3 nan older than light from the home. However, since our clocks don't read fractions of a nan, this accuracy is sufficient. If greater accuracy is demanded, then build the road 120 000 feet away and snap the photo 120 000 nans later.) If this same camera snaps a second photo 150 nans later (i.e., 12 150 nans after Rosemary leaves her home), that photo will be very much like the bottom sketch in figure 8.1.

In addition, station a photographer on a rocket sled traveling this new road at $V = \frac{3}{5}c$ exactly abreast of Rosemary. This photographer snaps photos 12 000 nans and 12 120 nans after Rosemary leaves her home (times ticked off by the rocket photographer's wristwatch), and these two photos are very much like the two sketches in figure 8.2. Of course, the rocket photographer will not be at the same place as the Earth-bound photographer when the photos are snapped.*

Problem-Solving Tips

Are there any tips for effective understanding that can be gleaned from the deli problem?

1. Keep your reference frame in mind. In the Earth's frame, Rosemary is moving and Rosemary's clock ticks slowly. In Rosemary's frame, the deli is moving and the deli clock ticks slowly. There is no single answer to the question "Does the deli clock tick slower than Rosemary's clock?" because the answer is no in the Earth's frame and yes in Rosemary's frame.

2. T versus T_0. In the same way, there's no single answer to the question "Does T_0 represent the time ticked by Rosemary's clock or by the deli clock?" The symbol T_0 represents the time ticked off by a single moving clock, so in the Earth's frame T_0 represents the time ticked off by Rosemary's clock, while in Rosemary's frame T_0 represents the time ticked off by the deli clock.

3. Sketch. Draw sketches for each problem. Make them big, so that you can see what's going on. Draw at least two sketches, one from each reference frame. But because two events that are simultaneous in one reference frame might not be simultaneous in another, you might need to draw one sketch from one frame and two sketches from another frame.

---

*Can you show that, in the Earth's frame, the rocket photographer snaps his first photo when he is 9 000 feet to the right of the Earth-bound photographer?

4. What you see is not what is happening. It takes some time for light to move from one place to another, so what Aaron sees is what happened at some time in the past, not what is happening now.

Solving a problem is like writing an essay. You write an essay by developing an overall structure for the argument and then fleshing out the details. You don't write even the first sentence until you have some idea of where you're going. And you certainly don't write an essay by paging through a book (the dictionary?) expecting to find the proper formula for you to "plug in."

Q: Why should I care about this stuff? The effects are only noticeable when traveling near the speed of light, and I'm never going to move that fast.
A: It is true that you'll never be moving near the speed of light relative to the Earth. Still, there are four reasons why you should be interested in relativity:

1. While humans don't move near the speed of light, there are things we care about that do. For example, cancer patients are often treated with radiation therapy from machines called linear accelerators. These machines move electrons at speeds of about 0.9997 c. The electrons are either directed at the tumor, or else used to make X rays that are directed at the tumor. In either case, if the machine designers used commonsense notions of space and time, rather than the correct relativistic notions, the machines simply would not work.

2. If high precision is needed, then relativistic effects need to be accounted for even if speeds are much less than light speed. For example, the Global Positioning System enables anyone with a GPS receiver to determine his or her position (latitude and longitude and elevation) to an accuracy of about 30 feet. The system has innumerable uses for truckers, hikers, miners, warfighters, and others. (Most cell phones sold after 2005 contain GPS receivers so that emergency calls to 911 can be tracked to their place of origin.) The system works because 24 satellites, each with a highly accurate atomic clock, are constantly broadcasting their positions and the time. A GPS receiver in contact with, for example, four satellites will receive four signals. Each signal has a different time stamp, because the radio signal from a more distant satellite takes a longer time to reach the receiver, so it bears an earlier time stamp. A computer within the receiver detangles all this information and uses it to find the receiver's position.

There are a lot of details concerning the GPS system (some of which are military secrets) but one thing's clear just from inspection: If your position is to be located with an accuracy of 30 feet or better, the clock readings must be accurate to 30 nans or better. That's 30 billionths of a second. On page 34 we calculated that each hour, a space shuttle clock loses 1.3 millionth of a second relative to an Earth clock. Back then we remarked that such a small discrepancy is hard to measure. But this discrepancy is 40 times larger than the accuracy demanded by a GPS clock. When accuracy demands are this stringent, relativistic effects must be accounted for—and indeed, relativity is incorporated throughout the workings of the GPS. For example, the satellite clocks, when manufactured at rest on the surface of the Earth, are adjusted to tick at a slightly different rate from Earth-bound clocks, so that when they're moving up in orbit they will tick at the same rate as Earth-bound clocks.*

3. Even if there were no practical applications of relativity, it would still be worth studying as a demonstration of the use of reason and observation combined to form a powerful tool for penetrating the unknown, even when the unknown violates common sense.

4. Finally, relativity is not just unexpected and bizarre but also beautiful, and demonstrates the surprising and satisfying harmony of this universe, our home.

PROBLEMS

8.1 ✦ **A trip to the grocery store.** Tabetha needs to pick up bananas from a grocery store located 1200 feet from her home. She travels there at a speed of $V = \frac{4}{5}c$. As she starts her journey, her wristwatch, her home's wall clock, and the grocery store's wall clock all read noon. (Tip: It will help to make four sketches: in the Earth's frame, leaving home and arriving at the store; in Tabetha's frame, leaving home and arriving at the store.)

    a. How much time elapses in the Earth's frame during her journey?

    b. How much time does Tabetha's wristwatch tick off during this journey?

    c. In Tabetha's frame, the grocery store moves toward her. How far must it travel to reach her?

    d. Using your result for part (c) and the definition

$$\text{speed} = \frac{\text{distance traveled}}{\text{time elapsed}},$$

find out how much time it takes for the grocery store to reach Tabetha. Compare with your result from part (b).

---

*If special-relativistic time dilation were the only effect in play, satellite clocks would be manufactured to tick quickly, so that when they move in orbit, time dilation would make them tick at the same rate as Earth-bound clocks. In fact, as we will see on page 154, the additional effect of gravitational time dilation also needs to be taken into account.

e. In Tabetha's frame, the grocery store's wall clock is moving, so it ticks slowly. How much time does it tick off during its journey to reach Tabetha?

f. When Tabetha reaches the grocery store, its wall clock reads the time you found in part (a). How can it read this time when it has ticked off only the smaller amount of time you determined in part (e)?

8.2 ✦ Two fast women. Veronica runs to the right past Ivan at speed $V = \frac{3}{5}c$. Veronica's sister Denise runs to the right past Ivan still faster, at speed $V = \frac{4}{5}c$ (in Ivan's frame). Ivan places two time bombs by the side of the road: one is 15 feet to the right of the other. The left-hand bomb explodes when its clock reads noon. The right-hand bomb explodes when its clock reads 9 nans after noon. How much time elapses between these two explosions in Veronica's frame? In Denise's frame? Sketch these two events using five pictures: two in Ivan's frame, one in Veronica's, and two in Denise's.

8.3 To see and to observe. The time is 130 nans after noon and Cynthia wishes to make an observation of the situation within 40 feet of where she's standing. At what time will she have all the information she needs to make this observation?

8.4 Racetrack. Alia finds a very long, straight racetrack, with two clocks mounted at either end of a 90-foot-long stretch near its middle. She happens to be carrying a 90-foot-long pole with a clock dangling from either end. (The two lengths given are rest lengths.) Alia, carrying her pole horizontally, runs to the right along the racetrack at speed $V = \frac{4}{5}c$. The Earth's clocks are synchronized in the Earth's frame, and Alia's clocks are synchronized in Alia's frame. As Alia's right-hand clock passes the Earth's left-hand clock, those two clocks both read noon. Here is the situation from the Earth's frame:

Racetrack from the Earth's frame

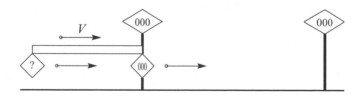

a. What is the length of Alia's pole in the Earth's frame?

b. In the figure above, the reading on Alia's left-hand clock is obscured by a question mark. What does it read?

c. As these two clocks line up and read noon, the situation in the Earth's frame is sketched above. Sketch the situation in Alia's frame, showing the reading on each clock and the two relevant lengths. (This is a simple problem, and *no calculation is required.*)

d. When Alia's right-hand clock passes the Earth's right-hand clock, what is the reading on the Earth's right-hand clock?

e. When Alia's right-hand clock passes the Earth's right-hand clock, what is the reading on Alia's right-hand clock?

f. Sketch the situation in both frames when Alia's right-hand clock passes the Earth's right-hand clock.

REFERENCES

The chapter epigraph is from

Tony Hillerman, *Listening Woman* (New York: Harper and Row, 1978), p. 34.

The use of relativity within the Global Positioning System is discussed in

Neil Ashby, "Relativity and the Global Positioning System," *Physics Today*, 55 (May 2002) 41–47.

One fascinating episode related in this article follows: "Nowadays the rate of every orbiting GPS clock is adjusted by this 'factory offset' before launch. But before the first GPS satellite was launched in 1977, although it was recognized that orbiting clocks would require such a relativistic offset, there was uncertainty as to its magnitude, and even its sign. So correcting frequency synthesizers were built into the clocks, spanning a large enough range around the nominal 10.23 MHz clock frequency to encompass all possibilities. After the satellite's cesium atomic clock was turned on, it was operated for three weeks to measure its rate. The frequency shift measured during this initial period was found to be 4.425 parts per ten billion, agreeing with the relativistic calculation to better than 1%." Details concerning this episode are reported in

J. A. Buisson, R. L. Easton, and T. B. McCaskill, "Initial results of the NAVSTAR GPS NTS-2 Satellite," pp. 177–99 in *Proceedings of the Ninth Annual Precise Time and Time Interval Conference, 1977* (NASA Technical Memorandum 78104),

which is reproduced at www.leapsecond.com/history/Ashby-Relativity.htm.

CHAPTER NINE

He Says, She Says

Time Dilation: Can they ever agree?

Veronica speeds past Ivan. Ivan says that Veronica's clocks tick slowly. Veronica says that Ivan's clocks tick slowly. Isn't this a contradiction?

It would be if time dilation were the only effect at work. But this section shows that because the full package of three effects—time dilation, length contraction, and the relativity of synchronization—are all at work, there is no contradiction. (There *is*, of course, a violation of common sense.)

To demonstrate the lack of contradiction, consider a concrete case in which Veronica runs past three of Ivan's clocks. As she runs by in a straight line, she compares her watch first to Ivan's left clock, then to Ivan's center clock, and finally to Ivan's right clock. We examine this situation first in Ivan's frame and then in Veronica's frame, and find that both examinations produce the same results for the actual clock comparisons. (For this demonstration I have chosen particular speeds and times to make the numbers come out neatly. If I had used different numbers, the numerical results would be different but the same consistency would appear at the end.)

Ivan has three clocks, uniformly spaced 30 feet apart and synchronized in his frame. Veronica, carrying a watch, travels to the right past Ivan at uniform speed $V = \frac{3}{5}c$. When Veronica passes in front of Ivan's first clock, all four clocks read zero nans in Ivan's frame.

Ivan's frame: start

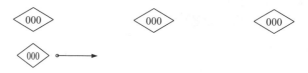

Some time later, Veronica passes in front of Ivan's middle clock. How much time later?

$$\text{time elapsed} = \frac{\text{distance traveled}}{\text{speed}} = \frac{30 \text{ ft}}{\frac{3}{5}c} = \frac{30 \text{ ft}}{\frac{3}{5}(1 \text{ ft/nan})} = 50 \text{ nan}.$$

So Ivan's three clocks all read 050 nans, but Veronica's clock has been ticking slowly by a factor of $\sqrt{1 - (V/c)^2} = \frac{4}{5}$, so it reads $\frac{4}{5}(50 \text{ nan}) = 40 \text{ nan}$.

Ivan's frame: middle

After 50 nans more have elapsed, Veronica passes in front of Ivan's last clock. Now the situation is

Ivan's frame: end

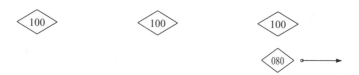

Now, what happens in Veronica's frame? There are three immediate differences. First, in Veronica's frame she is stationary and Ivan's clocks move left at speed $\frac{3}{5}c$. Second, Ivan's clocks are not separated by the length 30 feet: because of length contraction, they are separated by a shorter length, $\frac{4}{5}(30 \text{ ft}) = 24 \text{ ft}$. (The "length contraction factor" is the same as the "time dilation factor," namely $\sqrt{1 - (V/c)^2}$, which in this example is 4/5.) Third, in Veronica's frame Ivan's three clocks are not synchronized. His middle clock is to the rear of his left clock, so it is set *ahead* by a time

$$\frac{L_0 V}{c^2} = \frac{(30 \text{ ft})(\frac{3}{5}c)}{c^2} = \frac{(30 \text{ ft})\frac{3}{5}}{c} = \frac{(30 \text{ ft})\frac{3}{5}}{1 \text{ ft/nan}} = 18 \text{ nan.}$$

Thus, the starting situation in Veronica's frame is

Veronica's frame: start

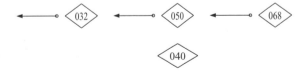

How long does it take for Ivan's middle clock to reach Veronica? (Remember that in this frame, Ivan's clock moves to reach Veronica, not vice versa.) The clock moves a (length-contracted) distance of 24 feet at a speed of $\frac{3}{5}c$, so the time required is

$$\text{time elapsed} = \frac{\text{distance traveled}}{\text{speed}} = \frac{24 \text{ ft}}{\frac{3}{5}c} = \frac{24 \text{ ft}}{\frac{3}{5}(1 \text{ ft/nan})} = 40 \text{ nan.}$$

This is the amount of time elapsing, and this is the amount of time ticked off by Veronica's clock. But Ivan's clocks are moving, so they tick slowly. Each of Ivan's three clocks ticks off, not 40 nans, but the smaller time

$$\frac{4}{5}(40 \text{ nan}) = 32 \text{ nan.}$$

So what is the situation in Veronica's frame when Ivan's middle clock speeds past her? It is

Veronica's frame: middle

The essential point: When the two clocks pass, Ivan's clock reads 50 nans and Veronica's clock reads 40 nans, and (compare the figure on page 76) *both parties agree on this fact!*

After 40 more nans elapse, Ivan's last clock passes Veronica, and all three of Ivan's clocks have ticked off an additional 32 nans.

Veronica's frame: end

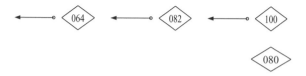

So, we return to our original question: Can Ivan and Veronica ever agree? Yes, they can agree on the essentials. They agree on the clock readings whenever Veronica passes in front of one of Ivan's clocks. They don't explain this in the same way: Ivan explains it by saying that Veronica's clocks tick slowly; Veronica explains it by saying that Ivan's clocks tick slowly and are out of synchronization. Although different, both explanations are correct.

There is nothing logically inconsistent about both clocks ticking slowly. A person standing in Los Angeles thinks (correctly!) that Tokyo is below his feet, while a person standing in Tokyo thinks (correctly!) that Los Angeles is below his feet. If the Earth were flat, this would be a contradiction. But the commonsense "flat Earth" idea is wrong. The Tokyo–Los Angeles situation is not a logical contradiction, and you're familiar with it, so it doesn't bother you.

In exactly the same way Ivan thinks (correctly!) that Veronica's clock ticks slowly, while Veronica thinks (correctly!) that Ivan's clock ticks slowly. If clocks synchronized in Ivan's frame were also synchronized in Veronica's frame, this would be a contradiction. But the commonsense "synchronized in all frames" idea is wrong. Although the Ivan-Veronica situation is not a logical contradiction, you're not familiar with it, so it probably does bother you.

PROBLEMS

9.1 Veronica in a truck. Make the situation completely symmetrical by placing Veronica in a 60-foot-long truck, moving at $V = \frac{3}{5}c$, with three clocks mounted on the truck. Sketch pictures like the six in this section, but now showing the readings on all six clocks.

9.2 Veronica peers ahead. At the instant when Veronica passes Ivan's middle clock, she looks ahead to see Ivan's right-hand clock. What clock reading does she see?

Measuring the Length of a Moving Rod

How can I find the length of a stationary rod? It's easy. I measure the distance from, say, a post to the left tip of the rod (5 feet in the situation below), then I measure the distance from that post to the right tip of the rod (11 feet), and finally I subtract those two distances (6 feet).

How can I find the length of a moving rod? In this case it's a little more complicated. If I measure the distance to the left tip first and then measure the distance to the right tip some time later, then I *don't* find the length of the rod through subtraction.

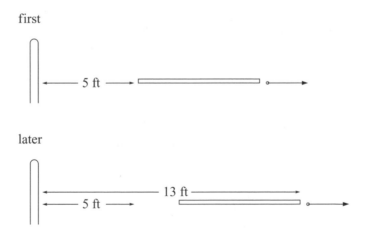

To find the correct length of a moving rod, I must add one more proviso to the prescription above: I must require that the two distance measurements be made simultaneously.

And there is the nub of the matter. Distances measured simultaneously in one reference frame might not be measured simultaneously in another reference frame, so two reference frames might legitimately disagree on the length of a moving rod.

The example below illustrates one way to put these measurement ideas into practice. Suppose a rod of rest length 5 feet speeds past us at $V = \frac{4}{5}c$.

We recruit a number of observers, line them up 1 foot apart, synchronize their watches, and instruct them: "Raise your hand if either rod tip is directly in front of you at noon." The figure shows the situation in our frame when all the watches read noon: two hands shoot up (the hands of observer #7 and of observer #10) and the distance between those hands is the length contracted amount 3 ft = $\frac{3}{5}$(5 ft).

Our frame

clock readings

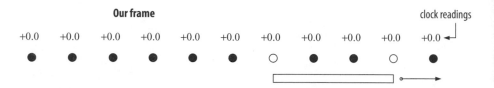

How does this measuring process happen in the rod's frame? There are three immediate differences. First, of course, the rod is stationary, and the observer recruits are moving left. Second, the observers are separated not by 1 foot but by the length-contracted distance of 0.6 foot. And third, adjacent watches are out of synch by $L_0V/c^2 = 0.8$ nan. Here's the situation in the rod's frame when the observer #10 raises his hand. (Observer #7 has not yet raised his hand, because his watch does not yet read noon.)

Rod's frame (first)

−7.2 −6.4 −5.6 −4.8 −4.0 −3.2 −2.4 −1.6 −0.8 +0.0 +0.8

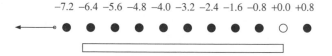

Now here's the situation exactly four nans later. The lined-up observers have moved a distance of $(\frac{4}{5}c)$(4 nan) = 3.2 ft. Each observer's watch has ticked off $\frac{3}{5}$(4 nan) = 2.4 nan. You can see that the watch of observer #7 reads noon when he's directly in front of the rod tip, and so he raises his hand.

Rod's frame (second)

−4.8 −4.0 −3.2 −2.4 −1.6 −0.8 +0.0 +0.8 +1.6 +2.4 +3.2

The two hands raise simultaneously in the Earth's frame but not in the rod's frame. This is why we need one figure to illustrate the action in the Earth's frame, and two figures to illustrate it in the rod's frame.

The rod says: Naturally those recruits got a short length. They didn't raise their hands simultaneously!

Once again, the two reference frames explain this phenomenon in different ways: The Earth-bound recruits say, "The moving rod is short." The rod says, "They didn't raise their hands simultaneously." But both frames agree on the essential fact that the two raised hands have exactly two observers between them.

As Galileo noted in his 1638 book, *Two New Sciences*, "Conclusions that are true may seem improbable at first glance, and yet when only some small thing is pointed out, they cast off their concealing cloaks and, thus naked and simple, gladly show off their secrets."

PROBLEM

9.3 ✦ American graffiti. Two commuters, Paul and Jennifer, stand on a railroad platform, each with a felt-tip pen and a wristwatch. Although dressed conservatively for their work-day business, each is at heart a secret anarchist and graffitist. They know that, at about 30 seconds before noon, the locomotive of a long express train will come roaring out of the west and rush by the platform at 3/5 the speed of light. The two agree that, exactly at noon, each will reach out with his or her felt-tip pen and mark the passing train. Jennifer stands 120 feet to the east of Paul.

In the reference frame of the train,

    a. Who marks first, Jennifer or Paul?

    b. For how much time does the train have only one mark on it? (Remember, in the train's frame both watches tick slowly.)

    c. In the train's frame, Paul travels west. How far does he travel while the train has only one mark on it?

    d. How far apart are the two marks on the train? Compare to $(120 \text{ ft})/\sqrt{1 - (V/c)^2}$.

(Clue: You will find this problem much easier if you familiarize yourself with the situation by making three sketches—one in the Earth's frame, corresponding to two in the train's frame.)

Synchronizing a Pair of Clocks

It's clear from the two examples presented in this chapter that the consistency of special relativity relies on the relativity of simultaneity. It is the keystone without which the whole scheme would collapse. Time dilation and length contraction by themselves generate contradictions. Only when the relativity of simultaneity is added to the package do the seeming contradictions vanish. But it is equally true that relativity of simultaneity is the hardest of these three principles to accept. So it is worth probing more deeply into this principle to see it in a different light.

How does one go about synchronizing two far-apart clocks? Here's one way: Suppose Darnell stands 30 feet from a wall clock, and he sees that the wall clock reads noon. Darnell *does not* set his wristwatch to noon. He knows that it took 30 nans for the light from the wall clock to reach him, so when he sees the wall clock read noon, he knows that the time is 30 nans after noon. In order to synchronize his wristwatch to the wall clock, when Darnell sees the wall clock reading noon, he sets his wristwatch to 30 nans after noon.

Darnell is riding in a train moving at $V = \frac{4}{5}c$ when he performs this synchronization procedure. Here are two pictures, in the train's reference frame, showing the light carrying the message "clock reads noon," first as it leaves the clock and second as it enters Darnell's eye.

Train's frame

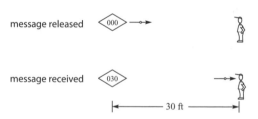

How does the synchronization process happen in the Earth's frame? First, the clock and Darnell are moving right. Second, the distance from the clock to Darnell is not 30 feet, but a length-contracted $\frac{3}{5}(30\ \text{ft}) = 18$ ft. Finally, the time required for the light to fly from the clock to Darnell is no longer 30 nans, but some time t_f that we need to determine.

During the synchronization process, the light travels a distance

$$ct_f = \tfrac{4}{5}ct_f + 18 \text{ ft}.$$

Solving this equation gives a flight time t_f of 90 nans.

While this is the amount of time elapsed, the moving wall clock ticks off a smaller amount of time, $\tfrac{3}{5}(90 \text{ nan}) = 54 \text{ nan}$. With this information we redraw the previous figure as

When Darnell sees that the wall clock reads noon, he sets his wristwatch to 30 nans after noon, and that's exactly what the wall clock reads in the train's frame. But in the Earth's frame, the wall clock reads 54 nans when Darnell sees the wall clock read noon and sets his wristwatch to 30 nans. So in the Earth's frame the wall clock (the rear clock) is set ahead of Darnell's wristwatch by 24 nans, which is $L_0 V/c^2$.

Here's a second way to synchronize two far-apart clocks. Suppose the conductor of a train wants to synchronize a clock at the train's tail and a clock at the train's nose. The conductor has two porters bring the two clocks together at the very center of the train, and there the two clocks are synchronized. The two porters then walk the two clocks back to their locations with equal speeds.

Examining this walk-back process in the train's frame, we see that the two clocks are both moving with respect to the train, so they're both ticking slowly. But because they're both moving at the same speed, they're both ticking with the same slowness and remain synchronized during the entire process.

Train's frame

In contrast, examine this process in the Earth's frame. The clock moving forward is moving *with* the train's motion, so it's moving very fast. The clock moving backward is moving *against* the train's motion, so it's moving not quite so fast. The clock moving forward ticks quite slowly, while the clock moving backward ticks not so slowly. When the synchronization process ceases, the clock that moved backward is set ahead of the clock that moved forward. The two clocks are synchronized in the train's frame, but in the Earth's frame the rear clock is set ahead.

Earth's frame

Q: You've given examples in which the three principles don't lead to inconsistencies, but of course you haven't considered all possible examples. Can you prove that special relativity will never lead to a contradiction in any situation?

A: Yes, I can. The proof consists of showing that the three principles are equivalent to a mathematical transformation called the Lorentz transformation and then showing that the set of all Lorentz transformations constitutes a mathematical "group." I could guide you through this process, but because you aren't familiar with mathematical transformations or mathematical groups, this proof will be unconvincing to you.* It is far more profitable to investigate a number of situations employing the three principles, showing how they seem to generate contradictions but in fact do not. And that's exactly what we're doing.

Q: Do you believe, then, that relativity provides the final word on the character of space and time?

A: Your question is an excellent one: just because a body of ideas is logically consistent doesn't necessarily make it correct. (For example, it would introduce no logical inconsistency if I were a millionaire, but in fact I am not.) To answer,

---

*Once, I presented this argument to a class, and one of my students told me that he could follow the algebra, but it wasn't a convincing argument because it was "just mathematical gobbledygook."

I'll begin with a story about a different field of science, namely geodesy, the science of the Earth's shape.

Primitive peoples thought that the Earth was flat—and as I look out of my second-floor office window, it certainly looks that way! (Understand that my office is located in Oberlin, Ohio.) But as early as 240 BC the Greek-Egyptian mathematician Eratosthenes considered the Earth to be a sphere and even devised a way to measure its radius. By 1687 Isaac Newton was treating the Earth as an oblate spheroid, that is, a ball bulging at the equator and flattened at the poles. The extent of this bulging is small—best current measurements are 6378.1 kilometers for the equatorial radius and 6356.8 kilometers for the polar radius, a difference of 0.3%—but it's perfectly clear that the Earth is *not* exactly a sphere.

The story doesn't end there: Just as the Earth lacks perfect spherical symmetry, so it lacks perfect axial symmetry. In 1828 the German mathematician Carl Friedrich Gauss discussed deviations from the oblate spheroid and introduced a shape for the Earth that's now called the *geoid*. The geoid is where sea level would be in the absence of winds, ocean currents, tides, and so forth. For example, the surface of the North Atlantic Ocean rises 85 meters above the oblate spheroid while the surface of the Indian Ocean plunges 106 meters below.

Geodesy continues to be an active and interesting field of research. In March 2002 two satellites were launched to perform the Gravity Recovery and Climate Experiment (GRACE), which is even now refining our measurements of the geoid. As recently as 7 January 2005 Minkang Chang and Bryon D. Tapley announced that they had used 28 years' worth of satellite laser ranging data to find changes in the Earth's shape and speculated that those changes were linked to El Niño climate oscillations. The science of geodesy is even today rife with controversy and open questions, as all active fields of science are and ought to be.

In fact, no one will ever have a perfect picture for the shape of the Earth because that shape changes every time a bulldozer sets to work and every time I overturn a shovelful of dirt in my garden! This doesn't mean that the science of the Earth's shape is all for nothing. Improvements in geodesy lead to improvements in mapping, air transportation, and mineral exploration and to improved calculations for the trajectories of missiles.

And while the flat-Earth approximation is wrong in detail, it is still quite appropriate for many applications. If I were to drive from 1617 183rd Street in Homewood, Illinois, to 301 North Ridgeland Avenue in Oak Park, Illinois, I would navigate using a flat map—the one titled "Chicago and Vicinity"—despite the fact that the Earth is not flat. All buildings are designed and constructed using the flat-Earth approximation.

Do I believe that the most recently published geoid provides the final word on the Earth's shape? No. On the contrary I'm sure that it isn't the final word. But it's closer to the truth than the common-sense flat-Earth approximation. (Or the sphere approximation, or the oblate-spheroid approximation.)

Do I believe that relativity provides the final word on the character of space and time? No. On the contrary I suspect that it isn't the final word. But it's closer to the truth than the commonsense classical approximation.

PROBLEMS

9.4◆ A chorus line. Each girl in a chorus line kicks her leg at the same time so that a photographer in the audience can take a snapshot in which all the legs are raised simultaneously. A car travels to the right at high speed. In the car's reference frame the girls do not kick simultaneously. Which girl kicks first, the one on the far right or the one on the far left? Will a photographer in the car need one snapshot, or two snapshots, or a video camera?

9.5 Relativity of synchronization rederived. Carry out the same reasoning that we used with Darnell synchronizing his watch, but (a) call the distance from the wall clock to Darnell L_0 instead of 30 feet, and (b) call the speed of the train V instead of $\frac{4}{5}c$. Use time dilation and length contraction to show that if two clocks are synchronized in the train's frame, then in the Earth's frame the rear clock is set ahead by $L_0 V/c^2$. (This constitutes a second derivation of the relativity of synchronization, which complements the derivation presented in chapter 7.)

9.6 Heartbeat. Veronica speeds past us in a railroad car going so fast that, because of time dilation, her heart beats not once a second but only once in ten seconds. Normally such a slow heart beat would be a cause for medical alarm: not enough oxygenated blood would be reaching her cells. Yet Veronica seems perfectly healthy. Why don't we need to worry about her slow heart beat?

9.7 Racetrack. Ivan finds a very long, straight racetrack, and he mounts a clock near its middle. Veronica places two clocks at either end of a 30-foot-long horizontal pole and runs to the right along the racetrack at speed $V = \frac{3}{5}c$. Veronica's clocks are synchronized in Veronica's frame. As Veronica's right clock passes Ivan's clock, those two clocks both read noon. The object of this problem is to find all the unknown clock readings in the figure.

Ivan's frame

first:

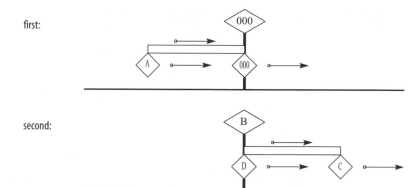

second:

Veronica's frame

first:

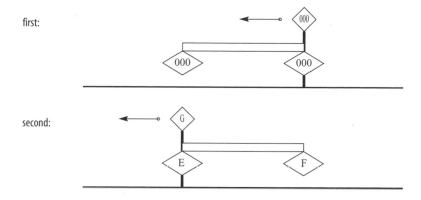

second:

a. How long is Veronica's pole in Ivan's frame?

b. What is clock reading A?

c. What are clock readings B, C, and D?

d. In Veronica's frame, Ivan's clock is moving to the left. How much time does it take for that clock to reach the left end of Veronica's pole?

e. What are clock readings E and F?

f. What is clock reading G?

g. Compare clock readings G and E to clock readings B and D.

CHAPTER TEN

Speed Limits

In one reference frame, two events happen in one sequence. In a different reference frame, those two events can happen in the opposite sequence. (See problem 8.2 on page 73 and the discussion on page 44.) This is unexpected, odd, and outside of our daily experience, but it's not a paradox: it's not logically inconsistent. We can live with the unexpected and with the unusual and with violations of common sense. (These happen every day in politics.) But we can't live with a logical inconsistency—that is, with a single complete statement's being both true and false.

So it's disturbing to see a potential paradox arise when the two events are related by cause and effect. Suppose one event causes another: for example, a snowball is tossed, and some time later that snowball splatters all over a wall. Could it be that in some reference frame these two events occur in the opposite sequence—that the snowball splatters first, and some time later the snowball is tossed? This certainly seems *wrong*: it seems that a cause should have to precede its effect in *all* reference frames. On the other hand, this book has introduced you to all sorts of things that seem wrong. Our strategy in this chapter will be to tentatively assume that cause precedes effect in all reference frames and then see where that assumption takes us. After examining the consequences, we'll go back and see whether our assumption still seems justified.

For concreteness, imagine the following situation in the Earth's reference frame. A snowball is tossed at time t_1 from the left and then splatters at time t_2 all over a wall on the right. The tosser and the wall are separated by a distance L_0. To help us keep track of time, we can imagine two clocks, anchored to the Earth, one located at the tosser and one at the wall.

Earth's frame

toss:

splatter:

The time elapsed during this process is $t_2 - t_1$, and the snowball travels at speed

$$v_b = \frac{L_0}{t_2 - t_1}.$$

How does this whole process transpire as observed from Veronica's frame, traveling to the right at speed V? In this frame (1) the tosser and wall move to the left, (2) the tosser and the wall are closer (length contraction), (3) the wall clock is set ahead of the tosser's clock (relativity of synchronization), and (4) both of the moving clocks tick slowly.

Veronica's frame

toss:

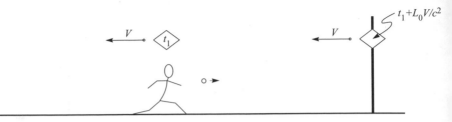

splatter:

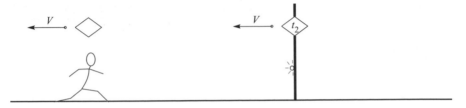

During this process, the clock at the wall ticks off time

$$t_2 - (t_1 + L_0 V/c^2).$$

But that's not how much time has elapsed, because the moving wall clock ticks slowly. The amount of time elapsed is a longer time

$$\frac{t_2 - t_1 - L_0 V/c^2}{\sqrt{1 - (V/c)^2}}.$$

Now we apply our assumption: that the time elapsed has to be positive in all reference frames:

$$\text{time elapsed} > 0$$

$$\frac{t_2 - t_1 - L_0 V/c^2}{\sqrt{1 - (V/c)^2}} > 0$$

$$t_2 - t_1 > L_0 V/c^2$$

$$1 > \left(\frac{L_0}{t_2 - t_1}\right) V/c^2$$

$$1 > \left(\frac{v_b}{c}\right)\left(\frac{V}{c}\right)$$

or, finally,

$$\left(\frac{v_b}{c}\right)\left(\frac{V}{c}\right) < 1.$$

This relation holds for all frames and for all causal signals: not just for the speed of a snowball, but also for the speed of a tomato, the speed of a baseball, the speed of sound, the speed of bullets, the speed of telephone signals, and the speed of light. Since it holds for all causal signals, it holds for a burst of light moving at $v_b = c$. Thus, for any inertial reference frame,

$V < c.$

No inertial reference frame can travel as fast as or faster than the speed of light.

> Q: I can't accept this! A reference frame is a purely conceptual device, and nature can't put a limit on purely conceptual entities. In my math class, I solve problems all the time about the populations of nations that don't exist, the heights of buildings that can't be built, and the trajectories of "point particles" that move in a "perfect vacuum" on a "flat Earth."
>
> A: A reference frame is *not* a purely conceptual entity. In order to have a reference frame, you need clocks, and measuring rods, and synchronization protocols, and recruits willing to stand in line and raise their hands. These are physical (not conceptual) things, so it's perfectly natural that they obey physical limits.
>
> Q: But what about a reference frame moving *at* the speed of light? You've already shown us how to make a clock out of light. Maybe we can make rods and such out of light too.
>
> A: You can't make a clock out of pure light—of light plus mirrors, yes, but not of pure light.

Can we find a similar speed limit on causal signals? Suppose an inertial reference frame moves at a speed just below the speed of light:

$V = c - \text{tiny},$

where "tiny" represents any positive speed as small as can be, but always bigger than zero. Then

$$\left(\frac{v_b}{c}\right)\left(\frac{V}{c}\right) < 1$$

$$\left(\frac{v_b}{c}\right)\left(\frac{c - \text{tiny}}{c}\right) < 1$$

$$\left(\frac{v_b}{c}\right)(1 - \text{tinier}) < 1$$

$$\frac{v_b}{c} < \frac{1}{1 - \text{tinier}}$$

$$\frac{v_b}{c} < 1 + \text{tinier}'$$

$$\frac{v_b}{c} \leq 1$$

$$v_b \leq c.$$

So, if our causality assumption is correct, then there is indeed a speed limit on causal signals: in an inertial frame no causal signal—no information—can travel faster than the speed of light.

Has this prediction been verified? Well, sending information is not just fun to speculate about, it's a multi-billion-dollar industry. The telecommunications industry, the television industry, the Internet industry, the computer industry, the courier industry: all, ultimately, are about sending information from place to place as quickly as possible. These industries work tirelessly to find ways of sending information quickly, but none has figured out a way of sending it faster than the speed of light.

Anything else? At the CERN laboratory near Geneva, Switzerland, scientists have figured out how to push a single electron so hard and so many times that, if commonsense notions of space and time were correct, the electron would be traveling 904 times the speed of light. But in fact no electron at CERN has traveled as fast as or faster than light. (The maximum speed achieved so far is 0.999 999 999 997 c.)

So despite the enormous monetary and scientific rewards that would be showered upon anyone sending a causal signal faster than light, no one has ever been able to do so—good evidence in support of both relativity and our causality assumption.

Q: Is it possible that, using future technology, information will someday be able to travel faster than light?
A: As far as we know, this is not an issue with technology. It is a limitation due to the character of space and time. Analogy: We can easily travel a thousand miles on the surface of the Earth, but it is difficult to travel a thousand miles up and even harder to travel a thousand miles down. This is a limitation in technology, which

makes it difficult but not impossible to move into the third spatial dimension. But we cannot travel a thousand miles into the fourth spatial dimension, not because of limitations in technology, but because the fourth spatial dimension does not exist.

Q: This is very disappointing. I always want to travel farther and faster, but now you tell me it's impossible.

A: Surprisingly, the restriction on speed does not restrict how far you can travel in a given time. (See chapter 15, "Voyage to Spica.") There are no relativistic limits on how far you can travel. (There are, of course, financial limits, but you already know that.)

Q: You keep on talking about a speed limit for "causal signals." What about for noncausal signals?

A: Noncausal signals have no speed limit. For example, focus your mind on the page in front of you. Then change the point of your mind's focus to a warm Hawaiian beach.* The point of focus has moved instantly—far faster than the speed of light! However, unlike a snowball, or a bullet, or a burst of light, the point of your mind's focus cannot cause anything to happen. It's not a causal signal.

Q: I'm a ballet dancer, and I can pirouette around in a circle in one second. While I'm spinning, in my reference frame everything rotates about me. For example, in just one second the Sun makes an orbit around my body. Now, the Sun is 93 million miles away, so in my reference frame it moves 580 million miles (circumference = $2\pi \times$ radius) in 1 second. That's way faster than the speed of light!

A: While you're spinning, you're not in an inertial reference frame. If you try to pour coffee while pirouetting at a dance performance, the coffee will not pour into the cup below the pot but will stream out over your audience. You can't send a message or information by setting the Sun into orbit around your body in this way.

Q: When I bang my thumb with a hammer, it hurts at my brain instantly!

A: No. The speed of nerve signals, about 200 miles/hour, is very fast by human standards but far below light speed.

Q: I read in the newspaper that scientists had made light travel "faster than light" by sending it through a medium.

A: Not really. You know that light affects the properties of materials it strikes: if you lie out in the sunshine you grow warm and, eventually, tanned. The scientists you mention made a material that was transparent but turned opaque soon after

---

*If you happen to be reading this page on a warm Hawaiian beach, then shift your focus to the North Pole instead.

light struck it. (Like getting a *very* fast tan.) If you send a beam of light that is, say, 3 feet long into a pane made of this material, the first 2 inches will emerge, but then the pane turns opaque and the rest of the beam is absorbed. The midpoint of this light beam moved very quickly: indeed, the midpoint of the emerging beam left the front of the pane before the midpoint of the arriving beam reached the back of the pane! But no individual piece of light traveled faster than the speed of light, c. The midpoint is just a mathematical concept and is allowed to travel as fast as the point of your mind's focus. You can't send a message on the midpoint of a light beam.

Q: What about tachyons?

A: Tachyons are hypothetical particles that always travel faster than light and can never slow down to light speed or below. If they exist, then an ordinary body is either always emitting tachyons, and emitting them in all directions, or else never emitting tachyons. It can't turn the emission on or off. Therefore, tachyons can't be used to send signals. (For example, you can send messages by switching a flashlight on and off in a Morse code pattern. But a flashlight that's always off sends no message. Similarly for a flashlight that's always on.) If tachyons existed, they would not violate relativity, but apparently they don't exist.* (Physicists have looked for them, and so far none have been found.)

Q: I read that in quantum mechanics, signals can move instantly.

A: Quantum mechanics is the physics of very small things, like atoms. It's very strange (far stranger than relativity), and the idea of cause and effect in quantum mechanics is subtle and richly textured. This subtlety has misled some into thinking that quantum mechanical signals move instantly, but the truth is that in quantum mechanics correlations jump instantly but information does not. (See the references for more discussion.)

Q: I'm riding in a railroad coach that's moving relative to the Earth at 10 miles/hour below the speed of light. I toss a tennis ball forward at 50 miles/hour relative to the train. Doesn't the tennis ball move, in the Earth's frame, 40 miles/hour faster than the speed of light?

A: Surprisingly, no! This is the topic of the next chapter.

Here's another scheme for moving a tennis ball faster than the speed of light which, like the "tennis ball in the railroad coach" scheme, doesn't work.

---

*Just because relativity allows something to exist doesn't mean that it has to exist. For example, nothing in relativity prohibits me from being dashingly handsome. But in fact I am not.

The conventional way of getting a tennis ball to move fast is to whack it with a tennis racket. You can make it go faster still by hitting it toward a partner, who whacks it again in the same direction. For a really fast tennis ball we could implement the following: Build a long, shallow trough to hold the tennis ball, hire a whole crew of tennis players to stand by the trough, and tell each athlete to speed up the ball with a whack whenever it goes by.* Pretty soon the ball will be going so fast that our athletes won't be able to see it, so we change their instructions to "each second, reach out into the space in front of you and swing your racket." Under these instructions, every second the ball gets one whack (although all but one of our athletes will reach out and swing through empty space). So it seems that every second the ball will add speed, and the speed will increase without limit.

But no. In our reference frame, the athletes are whacking once each second. Not so in the ball's reference frame. In the ball's frame, the athletes's clocks tick slowly, so the whacks come at intervals greater than 1 second. At first, this interval is just a fraction more than a second. By the time the tennis ball reaches 161 000 miles/second, it feels a whack once every 2 seconds. So it still picks up speed but not as readily as it did when it felt a whack once every second.[†] As the ball goes faster and faster, the whacks come slower and slower. It gets to be a minute between whacks, then an hour, then a year, then a century. Naturally, a ball that is being whacked once a century increases its speed at a very low rate. It doesn't pick up speed the way it did back when it was being whacked once a second.

So to the ball, the whacks are coming farther and farther apart. Our athletes are whacking as hard and as often as ever, but they're getting frustrated because they don't see results. They say the ball is becoming more and more resistant to picking up speed from a whack. The name for resistance to picking up speed from a whack is *inertia* or *mass*. (A lightweight tennis ball picks up a lot of speed from a single whack. A massive bowling ball picks up much less speed from an identical whack.) Our athletes say that as the ball travels closer and closer to the speed of light, its mass increases without limit.

One way of characterizing the ball's increased speed is through its increased energy. I haven't used the word *energy* before in this book because the physicist's term *energy* differs in important but subtle ways from the everyday term *energy*, and it would take many (rather tedious) pages to describe exactly how these two

---

*To keep our costs down, it might make sense to build the trough in the form of a circle and hire a limited number of athletes.

[†]In addition to this time dilation, in the ball's frame the athletes are standing closer together and not whacking simultaneously.

meanings differ. But I do want to show you one thing that summarizes our discussion: The ball explains its own behavior through a space-time effect: the whacks are coming slower. The athletes explain the ball's behavior through an increase in its mass. Scientists among the athletes, who have studied the physics definition of energy, find that the relation between increasing energy and increasing mass is

$$E = mc^2,$$

an equation that you might have encountered previously.

The interconversion of energy and mass implied in $E = mc^2$ is one of the most surprising and unexpected results of relativity, but also one of the most beautiful and useful. The Sun works by converting mass to energy through nuclear reactions. On page 71 I listed four reasons why you should study relativity, even though it's a small effect at everyday speeds. Here's a fifth reason: Sometimes small effects have large consequences. Relativity is responsible for the generation of all sunlight and all starlight. We are alive today only because relativity is true. If commonsense notions of space and time were correct, then the Sun would not work, and we would never have lived.

Q: Would it be fair to summarize all we've learned by saying "Nothing is absolute; it's all relative"?

A: No. This is a common misconception. For example, the author, diarist, and literary critic Anaïs Nin erroneously claimed that "the absolute became a fiction with Einstein."

Both in relativity and in common sense some things are relative and some are not. If you look at a building from the front or from the side, you see different things. The statement "Washington, DC, is 500 miles away" is true when made in Cleveland and false when made in San Francisco. The speed of a bird relative to the ground differs from the speed of that same bird relative to a runner. There are lots of things that depend upon point of view or frame of reference (i.e., that are "relative"). Einstein's relativity changes only which things depend on frame. In common sense, time interval doesn't depend upon frame, but the speed of light does. In the real world of relativity it's the other way around.

Observations from different reference frames can differ in unexpected ways, but they still represent different ways of observing the same things. If a vase falls to the floor and shatters, it shatters in all reference frames. This is an absolute, not a relative. Different frames might disagree on the details of falling, in the manner that we've seen in chapter 9, "He Says, She Says." But all frames will agree that the vase shatters.

PROBLEMS

10.1 Faster than light. I stand 10 feet from a white wall, holding a laser pointer. The pointer makes a red dot on the wall directly in front of me. Then with a quick flick of my wrist (lasting only 0.1 second) I twist the laser pointer by 45°, and the red dot moves 10 feet in 0.1 second. I repeat the experiment standing 20 feet from a white wall, and the red dot moves 20 feet in 0.1 second. Finally, I repeat the experiment standing 186 000 miles from a white wall, and the red dot moves 186 000 miles in 0.1 second. In other words, the red dot moves 10 times faster than light speed. Why doesn't this motion violate the speed limit derived in this chapter?

10.2 Paradox?

a. The year is 1492, and you are discussing with a friend the radical idea that the Earth is round. "This idea can't be correct," objects your friend, "because it contains a paradox. If it were true, then a traveler moving always due east would eventually arrive back at his starting point. Anyone can see that that's not possible!" Convince your friend that this paradox is not an internal inconsistency within the round-Earth idea, but an inconsistency between the round-Earth idea and the commonsense picture of the Earth as a plane, a picture which your friend has internalized so thoroughly that he can't recognize it as an approximation rather than the absolute truth.

b. The year is 1992, and you are discussing with a friend the radical ideas of relativity. "This idea can't be correct," objects your friend, "because it contains a paradox. If it were true, then two events could occur in one sequence in one reference frame and in the opposite sequence in a different reference frame. Anyone can see that that's not possible!" Convince your friend that this paradox is not an internal inconsistency within relativity, but an inconsistency between relativity and the commonsense picture of absolute space and time, a picture which your friend has internalized so thoroughly that he can't recognize it as an approximation rather than the absolute truth.

REFERENCES

"Faster-than-light" motion of light through a medium is described in

Lijun J. Wang, A. Kuzmich, and A. Dogariu, "Gain-assisted superluminal light propagation," *Nature*, 406 (20 July 2000) 277–79, and popularized through newspaper articles such as the one appearing in the *New York Times* of 30 May 2000. Tachyons are discussed in

E. Recami, "Classical tachyons and possible applications," *Rivista del Nuovo Cimento*, 9 (1986) 1–178.

Quantum mechanical faster-than-light effects are touched upon in

D. H. Freedman, "Faster than light fantasies," *Discover*, 19 (August 1998) 70–79,

but I (modestly) find the best treatment for a general audience to be

Daniel F. Styer, *The Strange World of Quantum Mechanics* (Cambridge: Cambridge University Press, 2000).

A high-accuracy experimental test of the famous energy-mass equation is

S. Rainville, J. K. Thompson, E. G. Myers, et al., "A direct test of $E = mc^2$," *Nature*, 438 (22 December 2005) 1096–97.

The misconception that "nothing is absolute" appears in, for example,

Anaïs Nin, *The Novel of the Future* (New York: Macmillan, 1968), p. 37.

CHAPTER ELEVEN

Speed Addition

A bird in flight is observed both from the Earth and from a moving car. Of course, these two reference frames find different speeds for the bird: for example, if both bird and car move right at 20 miles/hour relative to the Earth, then the bird's speed is 20 miles/hour in the Earth's frame and 0 miles/hour in the car's frame.

We examined this situation back on page 12, where we found that if

w_b represents the speed of the bird relative to the Earth,
v_b represents the speed of the bird relative to the car, and
V represents the speed of the car relative to the Earth,

then

$$w_b = v_b + V.$$

However, if you go back to the derivation of this formula on page 11, you'll see that the derivation made two assumptions: that a moving car has the same length as a stationary car, and that the time interval required for any process is the same in all reference frames. These assumptions, wonderfully in accord with common sense, are also false. What happens when we repeat the derivation, this time using the correct (but counterintuitive) ideas of time dilation, length contraction, and the relativity of simultaneity?

We'll do this derivation at the end of this chapter, and we'll show that in reality (that is, in relativity),

$$w_b = \frac{v_b + V}{1 + v_b V/c^2}.$$

This is the famous "Einstein speed addition formula." Before deriving it, we'll try it out with a few examples to see what it has to teach us.

First of all, what happens if we use ordinary speeds? Suppose I'm in a railroad coach traveling, relative to the Earth, at $V = 60$ miles/hour. I throw a tennis ball forward, relative to the coach, at $v_b = 20$ miles/hour. According to common sense, the speed of the ball relative to the Earth is $w_b = 80$ miles/hour. But what happens in truth? Plugging these numbers into the Einstein speed addition formula tells us that

$$w_b = 79.999\,999\,999\,999\,8 \text{ miles/hour.}$$

So for these ordinary speeds, the Einstein formula gives almost exactly the same result as common sense. The difference is so small that it couldn't be measured using typical speedometers.

But what happens if we use larger speeds? Suppose I'm in a railroad coach traveling, relative to the Earth, at $V = \frac{3}{4}c$. I throw a tennis ball forward, relative to the coach, at $v_b = \frac{3}{4}c$. According to common sense, the speed of the ball relative to the Earth is $w_b = 1\frac{1}{2}c$. This faster-than-light result is troubling, since we've seen that no ball can travel as fast as or faster than light. But plugging these numbers into the correct Einstein speed addition formula gives us the result that $w_b = \frac{24}{25}c$, which is safely below light speed.

Let's try the example mentioned on page 94. I'm in a railroad coach traveling, relative to the Earth, at $V = c - 10$ miles/hour. I throw a tennis ball forward, relative to the coach, at $v_b = 50$ miles/hour. According to common sense, the speed of the ball relative to the Earth is $w_b = c + 40$ miles/hour. But the Einstein formula gives the correct answer: $w_b = c - 9.999\,998\,5$ miles/hour.

What if the railroad coach is moving at any speed V, and v_b is the speed of light c? Then

$$w_b = \frac{c + V}{1 + cV/c^2} = \frac{c + V}{1 + V/c} = \frac{(c + V)c}{c + V} = c.$$

As must be the case for a correct formula, the Einstein formula shows that the speed of light is the same in all reference frames. But it shows more: Anything moving at the speed of light moves with the same speed in all reference frames. It's not light that's special, it's the speed of light.

The commonsense speed addition formula is highly accurate for speeds much less than the speed of light. But the Einstein speed addition formula works for all speeds, small and large, and smoothly interpolates between the "commonsense" result at small speeds and the "c = constant" result at high speeds. Sometimes people get the misimpression that all speeds behave in the commonsense manner until the speed of light is reached, at which point there is a discontinuous jump to the "constant in all reference frames" behavior. No. The transition is gradual and is governed by the Einstein formula.

The Einstein speed addition formula is subject to experimental test and has been tested many times. For example, there is a particle called a K^0 meson that decays into two pions. Whenever this happens with a stationary K^0 meson, the two daughter pions travel off in opposite directions at the speed of 0.85 c. Suppose a K^0 meson travels to the right at speed 0.72 c with respect to the Earth and decays with one of the daughter pions ejected to the right. Common sense says that, relative to the Earth, the daughter would travel at

$$w_b = v_b + V = 0.85\,c + 0.72\,c = 1.57\,c.$$

Einstein's formula says that

$$w_b = \frac{v_b + V}{1 + v_b V/c^2} = \frac{0.85\,c + 0.72\,c}{1 + (0.85)(0.72)} = 0.97\,c,$$

and this is in fact observed.

We will soon find that the speed addition formula is derived directly from the package of all three principles: time dilation, length contraction, and the relativity of simultaneity. Thus, an experiment testing the velocity addition formula is an indirect, but nevertheless very sensitive, test of all three of these principles.

PROBLEMS

11.1 ✦ Speed addition. Darnell stands in a railroad coach that moves at speed $\frac{1}{2}c$ relative to the Earth. He tosses a tennis ball forward at speed $\frac{1}{2}c$ relative to the train. If common sense were correct, the tennis ball would be moving at speed c relative to the Earth.

    a. How fast does the tennis ball really move relative to the Earth?

        Darnell stands in a railroad coach that moves at speed $\frac{1}{4}c$ relative to the Earth. He tosses a tennis ball forward at speed $\frac{3}{4}c$ relative to the train. If common sense were correct, the tennis ball would be moving at speed c relative to the Earth.

    b. How fast does the tennis ball really move relative to the Earth?

11.2 Two fast women. Veronica speeds to the right past Ivan at speed $\frac{3}{5}c$. Veronica's sister Denise speeds to the right past Ivan still faster, at speed $\frac{4}{5}c$ (in Ivan's frame). How fast does Denise move in Veronica's frame? How fast does Veronica move in Denise's frame?

11.3 High-speed approach. You have undoubtedly seen television sequences in which two lovers run across a flower-strewn meadow into each other's arms. Edwin runs right at speed $\frac{1}{2}c$. Carla runs left toward him at speed $\frac{3}{4}c$ (i.e., $w_b = -\frac{3}{4}c$). How fast does Carla move in Edwin's frame? (Caution! Don't try this at home. Running into anyone's arms while traveling at these speeds would be . . . what shall I say . . . unromantic.)

11.4 ✦ Tennis in a train. On pages 94 and 100 we discussed tossing a tennis ball within a railroad coach moving near light speed. We concluded that in the train's frame the tennis ball travels 50 miles/hour faster than the train, whereas in the Earth's frame the tennis ball travels 0.000 001 5 mile/hour faster than the train. Without using formulas, discuss the concepts that result in the ball moving so sluggishly relative to the train (from the Earth's frame).

Deriving the Speed Addition Formula: Numbers

Uncovering the Einstein speed addition formula is the most intricate derivation in this book. We approach the problem by first examining a specific situation with particular numbers, which will help to get the concepts straight, and then repeating the derivation for a general situation with symbols.

A ladybug flies within a railroad coach. We find its speed, in the coach's frame, by timing the flight with two clocks mounted in the railroad coach, located 60 feet apart. The ladybug passes the left clock at time 140 nans and the right clock at time 260 nans. Its speed is the distance traveled divided by the time elapsed. In detail:

1. The distance traveled by the ladybug is 60 feet.
2. The time elapsed is 260 nan − 140 nan = 120 nan.
3. So the speed of the ladybug is

$$v_b = \frac{\text{distance traveled}}{\text{time elapsed}} = \frac{60 \text{ ft}}{120 \text{ nan}} = \frac{1}{2}c.$$

Coach's frame

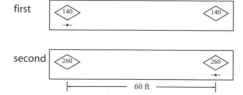

Earth's frame

first

second

The railroad coach travels at speed $V = \frac{3}{5}c$ relative to the Earth. What is the ladybug's speed, w_b, in the Earth's frame? If common sense were correct, it would be $\frac{1}{2}c + \frac{3}{5}c = 1\frac{1}{10}c$, faster than the speed of light.

What happens in the Earth's frame in reality?

1. The two clocks are separated by a smaller length, $\frac{4}{5}(60 \text{ ft}) = 48$ ft.

2. The right clock (front clock) reads behind* the left clock by $L_0 V/c^2 =$ $(60 \text{ ft})(\frac{3}{5})/(1 \text{ ft/nan}) = 36$ nan.

3. So when the ladybug passes the left clock, the right clock reads 140 nan − 36 nan = 104 nan.

4. The time ticked off (by the right clock) between the ladybug's passing the left clock and then passing the right clock is 260 nan − 104 nan = 156 nan.

5. But this moving clock ticks slowly, so the time elapsed is *not* 156 nan, but rather a longer time, $\frac{5}{4}(156 \text{ nan}) = 195$ nan.

6. During this time, the coach travels a distance $V \times (\text{time}) = (\frac{3}{5}c)(195 \text{ nan}) =$ 117 ft.

7. The distance traveled by the ladybug is the sum of items **6** and **1**: 117 ft + 48 ft = 165 ft.

8. The speed of the ladybug (in the Earth's frame) is obtained from items **7** and **5**:

$$w_b = \frac{\text{distance traveled}}{\text{time elapsed}} = \frac{165 \text{ ft}}{195 \text{ nan}} = \frac{11}{13}c.$$

Compare this speed with the result from the Einstein speed addition formula:

$$w_b = \frac{v_b + V}{1 + v_b V/c^2} = \frac{\frac{1}{2}c + \frac{3}{5}c}{1 + \frac{1}{2} \times \frac{3}{5}} = \frac{11}{13}c.$$

Things are looking good for Einstein's formula.

*Our familiar litany "the rear clock is set ahead by $L_0 V/c^2$" is equivalent to "the front clock is set behind by $L_0 V/c^2$."

Deriving the Speed Addition Formula: Symbols

I'm going to repeat this exact same calculation, this time using symbols rather than numbers to produce a formula that works for any speeds.

A ladybug flies within a railroad coach. We find its speed, in the coach's frame, by timing the flight with two clocks mounted in the railroad coach, located a distance L_0 apart. The ladybug passes the left clock at time t_1 and the right clock at time t_2. Its speed is the distance traveled divided by the time elapsed. In detail:

1. The distance traveled by the ladybug is L_0.
2. The time elapsed is $t_2 - t_1 = \Delta t$.
3. So the speed of the ladybug is

$$v_b = \frac{\text{distance traveled}}{\text{time elapsed}} = \frac{L_0}{\Delta t}.$$

Coach's frame

Earth's frame

The railroad coach travels at speed V relative to the Earth. What is the ladybug's speed, w_b, in the Earth's frame? If common sense were correct, it would be $v_b + V$.

What happens in the Earth's frame in reality?

1. The two clocks are separated by a smaller length, $\sqrt{1 - (V/c)^2}L_0$.
2. The right clock (front clock) reads behind the left clock by $L_0 V/c^2$.
3. So when the ladybug passes the left clock, the right clock reads $t_1 - L_0 V/c^2$.
4. The time ticked off (by the right clock) between the ladybug's passing the left clock and then passing the right clock is $t_2 - (t_1 - L_0 V/c^2) = \Delta t + L_0 V/c^2$.

5. But this moving clock ticks slowly, so the time elapsed is *not* the time ticked off, but rather a longer time, $(\Delta t + L_0 V/c^2)/\sqrt{1 - (V/c)^2}$.

6. During this time, the coach travels a distance $V \times$ (time) = $V(\Delta t + L_0 V/c^2)/\sqrt{1 - (V/c)^2}$.

7. The distance traveled by the ladybug is the sum of items **6** and **1**:

$$V(\Delta t + L_0 V/c^2)/\sqrt{1 - (V/c)^2} + \sqrt{1 - (V/c)^2}L_0$$
$$= [V(\Delta t + L_0 V/c^2) + (1 - (V/c)^2)L_0]/\sqrt{1 - (V/c)^2}$$
$$= [V\Delta t + L_0]/\sqrt{1 - (V/c)^2}.$$

8. The speed of the ladybug (in the Earth's frame) is obtained from items **7** and **5**:

$$w_b = \frac{\text{distance traveled}}{\text{time elapsed}}$$

$$= \frac{[V\Delta t + L_0]/\sqrt{1 - (V/c)^2}}{[\Delta t + L_0 V/c^2]/\sqrt{1 - (V/c)^2}}$$

$$= \frac{V\Delta t + L_0}{\Delta t + L_0 V/c^2} \qquad \text{Divide numerator and denominator by } \Delta t \ldots$$

$$= \frac{V + L_0/\Delta t}{1 + L_0 V/\Delta t c^2} \qquad \text{Remember that } v_b = L_0/\Delta t \ldots$$

$$= \frac{v_b + V}{1 + v_b V/c^2}.$$

There it is: We have derived the Einstein speed addition formula from time dilation, length contraction, and the relativity of simultaneity.

PROBLEMS

11.5 Logical progression. The logical progression of this book began with the constancy of the speed of light (for which there is ample experimental verification) and proceeded from there to time dilation, then to length contraction, then to the relativity of simultaneity, and finally to the Einstein speed addition formula. There are other, equally legitimate, logical pathways. For example, we could have begun with time dilation (for which there is ample experimental verification) and ended at the constancy of the speed of light. Outline this logical progression using four or fewer sentences.

11.6 ✦ Which clock set which way? The figure on page 67 shows a moving pair of clocks, with the right clock set *ahead of* the left clock. The figure on page 103 shows a moving pair of clocks, with the right clock set *behind* the left clock. Explain why in one circumstance it's the right clock that's set ahead whereas in the other circumstance it's the left clock that's set ahead.

REFERENCES

Experimental tests of the Einstein speed addition formula, such as those mentioned on page 101, are performed every day, almost every second, all over the world, in high-energy physics laboratories like CERN. However, the scientists doing these experiments call them not "tests of the speed addition formula," but instead "tests of relativistic momentum conservation." This is because the formula for relativistic momentum is dictated by the Einstein speed addition formula.

Examples of explicit momentum conservation tests are

C. M. Will, "Is momentum conserved? A test in the binary system PSR 1913+16," *The Astrophysical Journal Letters*, 393 (1992) L59–L61.

J. F. Bell and T. Damour, "A new test of conservation laws and Lorentz invariance in relativistic gravity," *Classical and Quantum Gravity*, 13 (1996) 3121–27.

But many routine measurements in nuclear and high-energy physics can be reinterpreted as tests of momentum conservation, and thus as tests of the Einstein speed addition formula. For example, the measurements by

E. N. Strait, D. M. Van Patter, W. W. Buechner, and A. Sperduto, "The reaction energies of light nuclei from magnetic analysis," *Physical Review*, 81 (1951) 747–60,

were reanalyzed in this way by

Edwin F. Taylor and John Archibald Wheeler, *Spacetime Physics*, 1st ed. (San Francisco: W. H. Freeman, 1966), pp. 123–33.

CHAPTER TWELVE

Rigidity, Straightness, and Strength

Rigidity. Here's a proposed technique for sending a signal faster than the speed of light—in fact, for sending it instantaneously: push the left end of a rod, and the right end moves at the same time! Well, not quite. When you push the left end of the rod, you move the first atom in a long chain of atoms that makes up the rod. A short time later, the first atom pushes the second, then the second pushes the third, and so forth. This push moves down the rod and reaches the end at a speed that is very fast by human standards,* so we don't notice it. But the speed is very slow compared to the speed of light. *There is no such thing as a perfectly rigid rod.*

To the four problem-solving tips listed on pages 70–71 I append one more:

5. Lack of rigidity. In relativity, all objects should be considered compressible and flexible like playdough.

Straightness. Ivan mounts a straight rod horizontally on three pegs, and he fits the base of each peg with a firecracker that can explode, causing its peg to crumble into bits. He arranges for the three firecrackers to explode simultaneously, and at the instant of the explosion the rod begins to fall down. The rod is always straight and always horizontal.

In Veronica's reference frame these three events are not simultaneous: First, the right peg crumbles and the right end of the rod begins to fall; second, the middle peg crumbles and the middle of the rod begins to fall; third, the left peg crumbles and the left end of the rod begins to fall. Between the first and second explosions the rod

*For a steel rod it moves at about 3 miles/second.

must be curved in Veronica's reference frame. *A rod that is straight in one reference frame may be curved in another.*

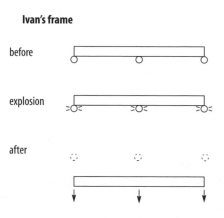

The rod is straight in Ivan's reference frame and curved in Veronica's. (The dashed circles represent the charred remains of exploded pegs.)

Q: Do you really believe this?

A: The word "believe" has a host of meanings. It is best used concerning matters of faith and ethics: "Do you believe in a trinitarian or a unitarian Christian God?" "Do you believe that children have a right to privacy?" These are matters of faith or opinion, and although they have been argued endlessly, at root they must be decided on the basis of belief and not on that of observation or experiment. In this sense of the word, scientists do not "believe" in relativity.

Instead, scientists "hold" relativity. We adopt relativity as the best explanation we currently have for the experiments we have so far performed. New experiments are being performed every day, and new explanations are being devised every day. Perhaps someday a reliable experiment inconsistent with relativity will be performed. When that day arrives, scientists are prepared to abandon relativity, just as Einstein was prepared to in 1921 (see page 19).

However, relativity is far more likely to be modified than completely over-turned. For example, suppose it were discovered that time intervals shorter than a trillionth of a second don't dilate in the same way that longer time intervals do. This would be an exciting discovery, and scientists would rush in to find out exactly how such ultrashort time intervals *do* behave, and to find the implications of this behavior for length and for synchronization. But these new experiments won't affect the already-performed experiments verifying time dilation for ordinary-sized time intervals. Whatever theory is ultimately developed for such ultrashort time intervals, it will need to merge smoothly into relativity theory as it is applied to longer and longer time intervals, just as relativity merges smoothly into common sense as it is applied to slower and slower reference frames.

In exactly this way, the flat-Earth model was abandoned for the spherical-Earth model, which was abandoned for the oblate-spheroidal-Earth model (see page 85). None of these is perfectly correct, but each is useful.

Strength. A 1-inch-long white line is chalked onto a bicycle wheel, parallel to the rim and perpendicular to the spokes. The wheel is mounted on a rack and spun up to very high speed. The white line moves parallel to the direction in which it points, so it is length-contracted. But the spokes move perpendicular to the direction in which they point, so they aren't length-contracted. (The spokes do decrease in thickness.) The white line is just one representative piece of the circumference of the wheel, so the whole circumference contracts. How can a wheel hold together with a contracted circumference and a noncontracted radius?*

The answer is that it can't. When a wheel rotates too fast, it breaks apart and flies into pieces. A wheel made of a weak material like wood will splinter at rather slow speeds; a wheel made of a strong material like steel will splinter at higher speeds; a wheel made of a very strong material like diamond will splinter at still higher speeds; but a wheel made up of *any* material will splinter at speeds much slower than speeds at which this relativistic effect becomes noticeable.† *There is no infinitely strong material.*

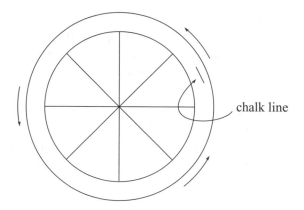

chalk line

The relativistic limits on the rigidity and strength of materials can be worked out quantitatively, and they are extreme. All known materials are much less rigid and much less strong than the limits allow.

There are no paradoxes within relativity, but there are contradictions between relativity and common sense. When you're working on a specific problem, it's easy to let common sense creep in, creating an apparent paradox. Unmasking such contradictions by ferreting out exactly where the common sense crept into your assumptions is a great way to learn relativity and to develop your relativistic intuition.

Q: How can we be sure of this?
A: We can't. All scientific statements are tentative; none are certain. The tenets and conclusions of relativity are among the most thoroughly tested in all of science, but they, like all scientific ideas, are based on experiments. No experiment is

---

*This question is named the Ehrenfest paradox.
†This is why I used a rocket sled rather than a rocket car in chapter 5, "The Great Race."

perfect, and not all potential experiments have been performed, so no scientific conclusion can be absolutely certain. Anyone worshiping at the altar of science is fooling himself.

One example: I emphasized on page 18 that the Michelson-Morley experiment has been performed many times under many circumstances and has always found the speed of light to be the same in all directions. But it's never been performed using an apparatus with arms only 1000 atoms long. It's never been performed in ultrahigh magnetic fields. Perhaps relativity doesn't hold under these circumstances.

Another example: Are you certain that you are reading a book entitled *Relativity for the Questioning Mind*? Or are you asleep and dreaming that you're reading a book entitled *Relativity for the Questioning Mind*? Any observation I point out suggesting that you're awake and reading (rather than asleep and dreaming about reading) you can dismiss by stating that you're dreaming, and that observations made within dreams are unreliable! No one can prove with certainty even such a simple fact as that she is awake. Instead of losing sleep over whether you're dreaming, you just get on with your life, realizing that probably you are awake. Scientists do the same thing. To insist on certainty is to insist that no progress will ever be made.

Some disciplines have authorities whose answers are always correct. If you think that a small home in the woods is a "cabin" while I think that it's a "kabin," then the dictionary shows that I'm wrong and there's nothing more to say. No amount of reasoning, observation, or experiment can change that. In spelling, the authority of the dictionary is absolute.

In contrast, any idea or theory in science can be overturned by reasoning, observation, and experiment. There is no book, guaranteed correct, containing the results of science. There never will be. In science, every observation can be elaborated upon, every measurement can be refined, every calculation can be improved, every generalization is tentative. Every idea can be disputed, tested, and refined. The only authority in science (the study of nature) is nature itself.

Some scientific verities (such as the idea that the Earth is round, the theory of relativity, atomic theory, and the theory of biological evolution) have been tested so many times in such a variety of circumstances that it's hard to imagine their being overturned. Nevertheless, in the year 1900 scientists felt the same way about commonsense verities of space and time, and (as we've seen) those verities *were* modified by Einstein!

In science, discovering that old verities are wrong is a source of pride and prizes. This contrasts with, say, theology, where anyone wishing to refine old verities may be considered a heretic. And it is precisely because science knows that it doesn't produce certainty, that the verities of science (while imperfect) are more reliable than the verities of theology.

I said on page 57 that science and religion are compatible. That's because religion is based on faith, while science is based on doubt. They appeal to different aspects of our humanity.

PROBLEM

12.1 Faster than light. Recall that the circumference and radius of a circle are related through

circumference = 2π × radius.

I grasp the butt of a foot ruler in one hand, then hold it horizontally pointing away from me. With a quick flick of my wrist (lasting only 1 second) I twist my hand by 90°, so the tip of the ruler moves $\pi/2$ feet in 1 second. (That is, about 1.6 feet in 1 second.) I repeat the experiment with a 2-foot-long stick, and the tip of the stick moves twice as far—about 3.2 feet—in 1 second. Finally I repeat the experiment with a stick 186 000 miles long, and the tip of the stick moves about 1.6 times light speed.

Obviously something's wrong with the analysis in the previous paragraph: We've seen that I can't make the tip of any stick move faster than light. What hidden (and incorrect) assumption did I make?

REFERENCES

The Ehrenfest paradox is discussed in detail in

Øvynd Grøn, "Space geometry in rotating reference frames: A historical appraisal," pp. 285–334 in *Relativity in Rotating Frames*, ed. Guido Rizzi and Matteo Luca Ruggiero (Dordrecht, the Netherlands: Kluwer Academic Publishers, 2004).

Grøn quotes Albert Einstein's 19 August 1919 letter to Joseph Petzold: "A rigid circular disk must break up if it is set into rotation, on account of the Lorentz contraction of the tangential fibres and the non-contraction of the radial ones. Similarly, a rigid disk in rotation must explode as a consequence of the inverse changes in length, if one attempts to bring it to the rest state."

Quantitative limits on rigidity and strength are worked out in

Wolfgang Rindler, *Introduction to Special Relativity* (Oxford: Clarendon Press, 1982).

In concert with the tentative and nondogmatic character of all scientific knowledge, scientists are searching for circumstances in which relativity fails. See

Alan Kostelecký, "The Search for Relativity Violations," *Scientific American*, 291 (September 2004) 92–101.

Jürgen Ehlers and Claus Lämmerzahl, eds., *Special Relativity: Will It Survive the Next 101 Years?* Lecture Notes in Physics, vol. 702 (Berlin: Springer, 2006).

CHAPTER THIRTEEN

The Twin Paradox

Veronica drives away from Ivan and then returns. She drives fast enough that her watch ticks at half the rate of Ivan's watch, so at the end of the trip Ivan's watch has ticked off 60 minutes and Veronica's watch has ticked off 30 minutes. This is strange and unfamiliar, but it's not a paradox.

Now consider the same trip from Veronica's point of view. To Veronica, Ivan moves away from her and then moves back, so *his* watch ticks slowly. Veronica thinks that while her watch ticks off 30 minutes his should have ticked off 15 minutes.

At the reunion at the end of the trip, Ivan's watch reads either 60 minutes or else 15 minutes. This is not a matter about which we can say "both are right." This is not just strange; this is a paradox. A situation that appears to be symmetrical is giving nonsymmetrical results. We have learned to live with the strange, but we cannot live with a logical contradiction.

The resolution of the paradox is that the situation only appears to be symmetrical. Halfway through her trip, Veronica changes her travel from "constant speed straight away from Ivan" to "constant speed straight toward Ivan." When Veronica is traveling in a constant speed in a straight line, her reference frame is just as good as Ivan's. But when she turns around, she is first thrown forward against her seatbelt as she brakes, then tossed to the outside of the curve as she turns, and finally thrown back into her seat as she accelerates again. At no point in the trip is Ivan thrown or tossed at all. During the turnaround the two reference frames are not "as good as each other," so the situation is not symmetrical and there's no logical contradiction. The correct analysis is the one performed entirely in Ivan's inertial reference frame: At the end of the trip, Ivan's watch has ticked off 60 minutes and Veronica's has ticked off 30 minutes.

Paradox lost.

This effect is called the "twin paradox" because if two twins go separate ways—one staying at home and the other traveling at high speed—then when they come back together, the traveling twin will be younger.

We've resolved the twin paradox: We know that Ivan's analysis is correct. But it's informative to see in detail how Veronica's analysis goes wrong, and that's what we'll do in the rest of this chapter. As usual, the situation is easier to understand if we pick a specific case. For variety, instead of studying a trip to the deli, which is finished in only nans, we'll consider a trip to the stars that requires years.

We will return to the twin paradox on page 156 after we have discussed accelerated motion. This second treatment will not invalidate anything we say here but will add the insight of an additional perspective.

Voyage to Sirius

Sirius is the brightest star in the night sky, a brilliant blue-white star located just below the constellation Orion. Any astronomy textbook will tell you that Sirius is located 8 light-years away from Earth.* This means that (in the Earth's frame) it takes 8 years for light, traveling at 186 000 miles/second, to fly from Sirius to Earth. So the distance from Earth to Sirius (in the Earth's frame) is

distance traveled

$$= \text{speed} \times \text{time elapsed}$$

$$= c \times (8 \text{ yr})$$

$$= \left(186\,000\,\frac{\text{mi}}{\text{sec}}\right) \times (8 \text{ yr}) \left(365\,\frac{\text{day}}{\text{yr}}\right) \left(24\,\frac{\text{hr}}{\text{day}}\right) \left(60\,\frac{\text{min}}{\text{hr}}\right) \left(60\,\frac{\text{sec}}{\text{min}}\right)$$

$$= 46\,925\,568\,000\,000 \text{ mi.}$$

This enormous number is so awkward to work with that we'll never use it again. We'll say instead that the distance to Sirius is $(8 \text{ yr})c$.

Veronica decides to take a trip from Earth to Sirius and back again, traveling at $V = \frac{4}{5}c$. Ivan chooses to remain at home on Earth.

Let's get two issues out of the way at the very start. First, the Earth and Sirius move relative to each other (if nothing else, due to the Earth's orbit around the Sun),

---

*The light-year is a unit of distance, not a unit of time.

but this relative motion is so slow relative to $\frac{4}{5}c$ that we can safely ignore it and consider Sirius to be at rest in the Earth's frame. Second, this relative motion means that the distance from Earth to Sirius changes with time, but these distance changes are so much smaller than 8 light-years that we can safely ignore them too.

The high points of Veronica's journey, as observed from the Earth's frame, are shown in figure 13.1. (The rectangular clocks display time measured in years.) You should be able to calculate all of these clock readings yourself: the only principles employed are the definition speed = distance/time and time dilation.

One more issue requires attention here. Veronica doesn't just step into her spaceship and then instantly move at $\frac{4}{5}c$, any more than you step into your car and then instantly move at 60 miles/hour. It takes some time (a "period of acceleration") for Veronica to get up to her cruising speed. Exactly how much time it takes will depend on the kind of spaceship Veronica uses, but let's say it's a week: it can't be instant, but the acceleration period can be small compared to the many years of total travel time. Similarly for the turnaround at Sirius: let's say it takes 2 weeks for her to slow down, turn around, and then get up to cruising speed for the return leg of journey. But since the clock readings shown in the figure are accurate only to the nearest tenth of a year anyway, these acceleration periods can safely be ignored.

The upshot is that at the end of the trip Ivan's clock has ticked off 20 years while Veronica's clock has ticked off 12 years. This applies not only to wristwatches, but also to biological clocks. Ivan will have aged 20 years and Veronica will have aged 12 years, so there will be more wrinkles on his face than on her face. Ivan explains this by saying that Veronica's moving clock ticks slowly. How does Veronica explain it?

We examine the outbound leg of the journey in Veronica's frame. As usual, changing from the Earth's frame to Veronica's frame involves four differences:

1. Instead of Veronica moving right, the Earth and Sirius move left.

2. The Earth and Sirius are closer (length contraction).

3. Clocks on the Earth and Sirius are not synchronized (relativity of synchronization).

4. The Earth and Sirius clocks tick slowly (time dilation).

These differences are exactly the same as those we examined in chapter 8, "The Case of the Hungry Traveler," so I'm not going to detail how I arrived at the numbers on line A in figure 13.2. You should do this for yourself. Be sure to notice the qualitative character of these results: the distance from Earth to Sirius is *shorter* in Veronica's frame. The Sirius clock is to the rear, so it is set *ahead*.

Earth's frame

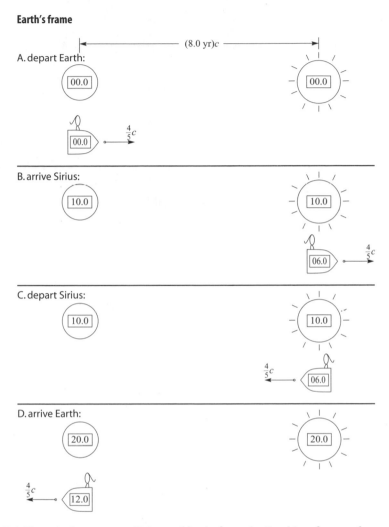

Fig. 13.1. Veronica's voyage to Sirius and back, from the Earth's reference frame.

How much time elapses before Sirius comes to meet Veronica? This is just

$$\text{time elapsed} = \frac{\text{distance traveled}}{\text{speed}} = \frac{(4.8 \text{ yr})c}{\frac{4}{5}c} = 6 \text{ yr},$$

so 6 years elapse and Veronica's clock ticks off 6 years. The Earth and Sirius clocks tick slowly, of course, so they tick off only $\frac{3}{5}(6 \text{ yr}) = 3.6 \text{ yr}$. These results are reflected in line B of figure 13.2.

Veronica's frames

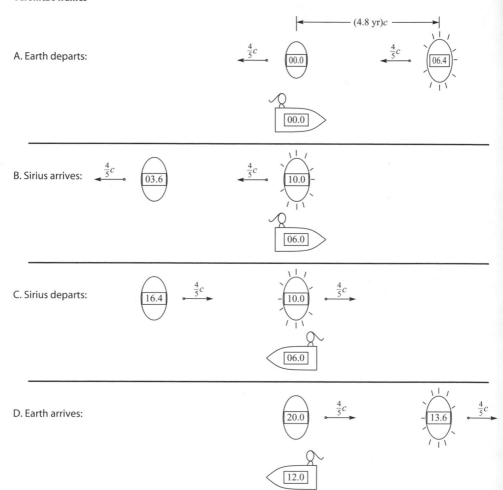

Fig. 13.2. Sirius's voyage to Veronica and back, from Veronica's two reference frames.

The next step, as described from the Earth's frame, is that Veronica slows down, turns around, and heads back to Earth. She's slipping out of one inertial frame and into another. But from Veronica's point of view, the difference is that the Earth and Sirius are no longer moving left; instead, they're moving right. On the outbound leg of the journey Earth is to the *front* of Sirius, so its clock is set *behind* by 6.4 years. On the return leg Earth is to the *rear* of Sirius, so its clock is set *ahead* by 6.4 years. Compare closely lines B and C in figure 13.2. During the turnaround, which requires

2 weeks for Veronica, the Earth clock jumps ahead by 12.8 years! This is the relativity of synchronization in action.

The return leg of the journey is like the outbound leg: Veronica's clock ticks off 6 years, and the Earth and Sirius clocks tick off 3.6 years. Veronica finds that when she returns home, her clocks have ticked off 12 years while Ivan's have ticked off 20 years.

Ivan explains this difference by saying that Veronica's clock ticks slowly. Veronica explains it by saying that Ivan's clock ticks slowly, except that during the turnaround (when his clock changed from front clock to rear clock) it ticked very rapidly indeed.

Q: So the resolution of the paradox hinges on the acceleration when Veronica turns around. What if she didn't make a drastic acceleration but instead took a circular flight at constant speed that gently looped back to her starting point?

A: We're not prepared to handle this problem in mathematical detail, but the idea should be straightforward: A circular flight at constant speed does involve acceleration because acceleration means a change in speed or direction of motion, and circular motion constantly changes direction. Indeed, throughout this whole trip Veronica will be tossed to the outside of the circle as she turns. In the out-and-back flight, there is a large acceleration lasting for a short amount of time. In the circular flight, there is a small acceleration lasting for a long amount of time. Ultimately, they have the same effect.

Q: If Veronica looks back at Earth, will she see Ivan age 12.8 years in 2 weeks? That would be gross!

A: No, because it takes time for the light from Ivan to reach Veronica. This topic is treated in problem 13.4, "Keeping in touch."

Q: Have any experiments been performed to test these conclusions about the twin paradox?

A: Yes. Many of the experiments discussed on page 35 test the twin paradox as well as simple time dilation. For example, the Navy antisubmarine flight involved one clock that stayed at the airport while another flew away and then returned. When the two clocks were reunited, the traveling clock had ticked off less time. (Remember that in Veronica's frame the explanation for the twin paradox involves both time dilation and the relativity of simultaneity, so these experiments test both principles at once.)

Q: I've been reading carefully, following the arguments critically, and I'm usually able to work the problems. But the whole subject of relativity still seems to me like a foreign land with inscrutable natives practicing incomprehensible customs. Will there ever come a time when the whole thing just "clicks" for me, and I understand it?

A: I can't tell what will happen to you, but I can tell what has happened to me: I started learning about relativity over 30 years ago, and it hasn't yet clicked for me. Every time I return to the subject, I find something new and unexpected and puzzling. For the first decade, this alarmed (and to some extent depressed) me. But I've learned to live with and even enjoy my ignorance. The things I find that are new and unexpected and puzzling are also stimulating and exciting. I would hate to ever feel that I had mastered relativity and that it would never again surprise or delight me.

The twin paradox offers a limited mechanism for time travel. Not travel backwards in time but, in a certain sense, travel forward in time. Do you want to know what life will be like on Earth in 300 years? Just take a round trip to Sirius, and if you go fast enough you will age 2 years during your excursion while 300 years pass on Earth. Of course, you can't go back to the friends you've left behind and tell them what's going to happen in 300 years; you will never see these friends again.

Given current technology this mechanism for time travel is grossly impractical, but it is nevertheless possible in principle. It is interesting to speculate about what would happen to human society if it ever did become practical. Suppose you suffer from an incurable disease that will kill you in 2 years. Should you jump 300 years into the future in hopes that someone will have found a cure for your disease? What about the risk that in 300 years human society will have entered a new "dark age"? How would you pay for your medical care when you arrive in the distant future? (You could try to use compound interest—but would any banker want to take you up on the deal?) Will medical research die out because all the difficult cases are just shoved into the future? Could prison sentences be replaced by "forward time jumps"? After a few hundred years, human society would be bombarded by travelers descending from out of the past. How would "stay at homes" feel about these relics from the past coming to live with them?

PROBLEMS

13.1 ✦ **Numbers.** Calculate the numbers shown in figures 13.1 and 13.2.

13.2 **The return of the hungry traveler.** Recall the case of the hungry traveler discussed in chapter 8. Suppose that when Rosemary reaches the deli, she instantly turns around and travels home at the same speed. Draw figures like figures 13.1 and 13.2 for this situation. How much does the home clock increment when Rosemary turns around?

13.3 Time travel. Ivan's doctor breaks the sad news that Ivan has been diagnosed with an incurable disease and has just over 2 years to live. "Medical researchers are working on a cure right now," the doctor explains, "and it will be available in twelve years. But I'm afraid that this will be too late to help you." Rather than give in to despair, Ivan jumps into a fast rocket ship and travels at high speed. When he returns to Earth he has aged 2 years and is near death. But 12 years have elapsed on Earth, the medical research has been successful, and Ivan is cured. How fast did Ivan travel? (Ignore the brief periods of acceleration.)

13.4 Keeping in touch. Before Veronica departs for Sirius at $V = \frac{4}{5}c$, she and Ivan agree to keep in touch during their long separation.

a. Ivan sends Veronica a radio message once a day. (Radio messages travel at the speed of light.) Because these messages have to catch up to Veronica, she does not receive them once a day. (This is not a relativistic effect; it happens in the commonsense world as well.) Show that the message Veronica receives as she turns around at Sirius was sent by Ivan 2 years after Veronica left Earth. Therefore, Veronica receives 1/10 of Ivan's messages on the outbound leg of her journey and 9/10 of them on the return leg.

b. Argue that Veronica receives these messages at regular intervals on the outbound leg and also at regular intervals on the return leg, but that the time between receptions is 9 times longer on the outbound leg than on the return leg.

c. Veronica sends Ivan a radio message once a day. Show that the message Veronica sends from Sirius reaches Ivan when his clock reads 18 years. Show also that during Ivan's first 18 years of separation, he receives—at regular intervals—the messages sent by Veronica on the outbound leg of her journey, and that during Ivan's final 2 years of separation, he receives, again at regular intervals, the messages sent by Veronica on the return leg of her journey. What is the relation between the interval of the first 18 years and the interval of the final 2 years?

d. During their separation Ivan trains his telescope on Veronica once a day. Show that during Ivan's first 18 years he sees the image in his telescope age (at a uniform rate) by 6 years, and that in Ivan's final 2 years he sees the image age (at a uniform rate) by 6 years again.

e. During their separation Veronica trains her telescope on Ivan once a day. Show that during the outbound leg she sees the image in her telescope age (at a uniform rate) by 2 years, and that during the return leg she sees the image age (at a uniform rate) by 18 years. Her telescope image does not show a sudden jump in Ivan's age.

13.5 Speedup. As she approaches Sirius, Veronica decides that she enjoys her life of travel and has no desire to return to Earth. So instead of slowing down and turning around at Sirius, she speeds up until she's moving away from Earth at $V = \frac{9}{10}c$. After she finishes her brief acceleration period, what is the time on Earth? If a baby was born on Earth 3 years after Veronica's departure, will Veronica see it born twice?

13.6 Twin paradox in fiction. The 1968 movie *Planet of the Apes* invokes the twin paradox: A spaceship crew traveling at high speed turns around and returns to Earth, only to find that much more time has elapsed on the Earth than has elapsed in their spaceship. Watch the movie and extract numerical data to find the speed of the spaceship. Did the film's director use special relativity correctly and consistently?

13.7 Fast forward. Write a short story implementing a trip to the future as discussed on page 120. As you learned in English composition class, your story will be more powerful and vigorous if you include specifics and details to back up the flow of your plot. Such specifics include character development, setting, and numerical information about the journey.* How fast and how far does the traveler move? How far into the future does he plunge? How much does the home calendar jump ahead when the traveler turns around?

*In his book *The Pine Barrens* (New York: Farrar, Straus & Giroux, 1967), John McPhee could have described a certain scene by writing, "There was a junk car by the side of the road." Instead, he made the image vivid and tangible by writing that the car "had ninety-three bullet holes in the driver's door alone."

CHAPTER FOURTEEN

The Pole in the Barn

M ost barns have two doors, one on the front and one on the back, so that you can pull a trailer into the barn, stop and unload it, and then drive out again without backing up. The farm where I grew up in Pennsylvania had a barn extending exactly 100 feet between its two doors.

One day a world-champion pole vaulter came to visit our farm. He carried his favorite pole which, by coincidence, was also exactly 100 feet long. The champion boasted that he was so fast that, even carrying his pole horizontally, he could run right through our barn at the speed of $V = \frac{4}{5}c$.

"At that speed," he assured my father, "my pole will be length-contracted until it's only $\frac{3}{5}(100 \text{ ft}) = 60 \text{ ft}$ long. I'll be able to fit it completely within your old barn! Look, if you don't believe it, put me to the test. Station one of your sons at the front door and the other one at the back door. Start with both doors closed, and open each one for just long enough to allow me through. You'll see. There will be a time when both doors are shut and my pole is completely enclosed within your barn."

My father was no dummy. He rubbed his chin and looked puzzled and thought-ful for a minute. "Okay, you're on," he told the pole vaulter. "I'll station my boys. But there's just one thing I don't understand: Sure, in the barn's frame your pole will be length-contracted. But in *your* frame the *barn* is moving. In *your* frame my barn is 60 feet wide and your pole is 100 feet long. How are you going to fit that long pole of yours into my stubby little barn?"

Now it was the champion's turn to be puzzled. In fact, he looked frightened and just a little greenish. I could see the sweat bead up on his forehead, and he lost his confident swagger. He wanted to bail out. My older brother went up and whispered a few words into the vaulter's ear. They huddled in quiet conversation for a few minutes, and then the vaulter regained most of his lost confidence. He carried out the feat flawlessly.

What did my brother tell the vaulter?

* * * * *

"You have to think about time and simultaneity also," my brother whispered, "not just about length."

"If I'm going to risk my life," the vaulter responded, "you'll have to be a lot more detailed than that."

"Fair enough," said my brother. "To make the situation concrete, let's suppose there are two clocks mounted on the barn, one at the front door and one at the back door. These clocks read time in nans. When the tip of the pole enters the front of the barn, both clocks read 0." With his finger, he sketched a figure in the farmyard dust.

Barn's frame: tip enters, front door opens

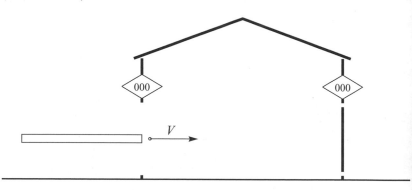

"Now, when does the butt of the pole enter the front door?" my brother continued. "It enters when the tip of the pole has moved 60 feet—that is, after this amount of time has passed:

$$\text{time elapsed} = \frac{\text{distance traveled}}{\text{speed}} = \frac{60 \text{ ft}}{\frac{4}{5} \text{ ft/nan}} = 75 \text{ nan.}$$

So here's the situation when the front door closes," he said, revising his sketch.

Barn's frame: butt enters, front door closes

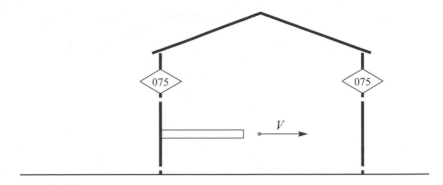

"Then the tip of the pole reaches the back door when the pole has traveled 40 more feet, requiring 50 more nans.

Barn's frame: tip exits, back door opens

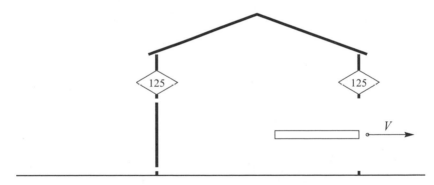

And finally the butt of the pole passes through the back door after the pole has traveled another 60 feet (requiring another 75 nans)."

Barn's frame: butt exits, back door closes

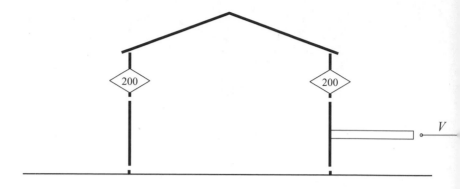

"Yes, I've got all that," whispered the worried vaulter, "but what about in *my* frame?"

"That's the neat part," replied my brother. "Those two barn clocks that were synchronized in the barn's frame. Are they synchronized in your frame?"

"No, of course not," said the vaulter. "The rear clock is set ahead of the front clock by

$$\frac{L_0 V}{c^2} = \frac{(100 \text{ ft})(\frac{4}{5})}{1 \text{ ft/nan}} = 80 \text{ nan."}$$

"Perfect," said my brother. "So in the vaulter's frame, when the tip of the pole enters the front door, the situation is this."

Vaulter's frame: tip enters, front door opens

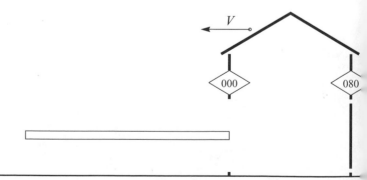

"I think I've got it," said the vaulter. "The next significant event is when the tip of my pole reaches the back door and the back door opens. This happens after the barn has moved 60 feet, which requires the passage of 75 nans. The situation then is this."

Vaulter's frame: tip exits, back door opens **[first try]**

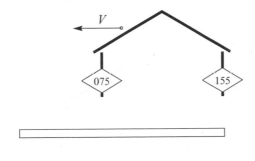

"But hold on," the vaulter said. "We've already seen that in the barn's frame, the pole tip exits the back door when the barn's back door clock reads 125 nans. In this picture, the back door clock reads 155 nans. That can't be right."

My brother looked concerned for a few seconds, then broke into a sly smile. "You're right that 75 nans have elapsed, but did those clocks tick off 75 nans?"

"No, no, of course not!" exclaimed the vaulter. "Those two moving clocks tick slowly! When 75 nans have elapsed, they've ticked off only $\frac{3}{5}(75 \text{ nan}) = 45$ nan. The situation really is this."

Vaulter's frame: tip exits, back door opens **[corrected]**

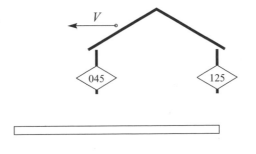

"Yep," agreed my brother, "if you apply only two of the three rules, you run into contradictions. That's just what happened on our first try, when we both forgot to apply time dilation."

"Okay, that should take care of all the problems," the vaulter suggested optimistically. "Next the barn moves 40 feet in 50 nans, during which time the moving clocks tick off 30 nans to give this situation.

Vaulter's frame: butt enters, front door closes

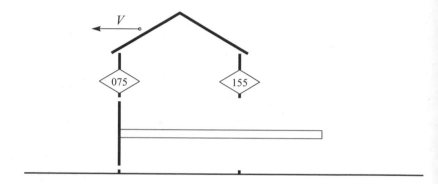

And then the barn moves 60 feet in 75 nans while the moving clocks tick off 45 more nans.

Vaulter's frame: butt exits, back door closes

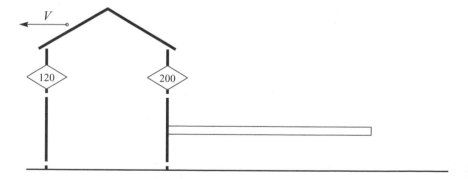

"So in the barn's frame," concluded the vaulter, "first the front door closes, and then, a little while later, the back door opens. During this interval my pole is entirely within the barn. But in the vaulter's frame, *my* frame . . ."

He trailed off, unwilling to say out loud the conclusion that his own calculations had shown him to be true. My brother finished for him. "In *your* frame, those two events happen in the opposite sequence. First the back door opens, then the front door closes. Your pole is never entirely within the barn."

"Really?"

"Sure. As my kid brother likes to say, 'Two events that occur in one sequence in one reference frame might occur in the opposite sequence in a different reference frame.'"

Armed with his new understanding, the vaulter moved back to allow himself a long run to get up to speed. He looked at the barn and hesitated once more. "Hold on," he said. "There aren't any clocks mounted on your barn. How can the argument work if the clocks aren't there?"

"The clocks," explained my brother, "were just scaffolding to propel the argument. Time passes whether or not human-made clocks are there to record its passing. Relativity actually deals with time and space, even though our formalism deals with clocks and measuring rods."

"Can I really believe this?" the vaulter asked meekly.

"Einstein's reasoning stands behind it," my brother replied. "If it weren't true, then the speed of light would be different in different reference frames."

PROBLEMS

14.1 Cloudburst. Suppose there's a brief cloudburst while (in the barn's frame) the pole is completely enclosed within the barn. Do the tips of the pole get wet or stay dry?

14.2 Fatal mistake. The pole vaulter grew to enjoy this trick, and he did it so many times that he got sloppy. One day he asked my brother and me to man the doors as usual, but he performed the stunt before I could walk around and station myself at the back door. As you can imagine, running into a thick barn door at speed $\frac{4}{5}c$ was, well, unhealthful. When the police arrived at our farm to investigate the vaulter's suspicious death, they found many things that puzzled them, but one question stood out: When the vaulter died, was the front barn door open or closed?

14.3 Electrical connection. Suppose that the back door is not opened by someone standing at the back door, but instead opens upon receiving an electrical signal generated when the front door closes. In this situation the back door opening is caused by the front door closing, so (causality assumption, page 88) the back door opening has to follow the front door closing in all reference frames. What happens in this situation?

14.4 Chain connection. Suppose that the back door is not opened by someone standing at the back door, but instead there is a chain contraption between the two doors so that closing the front door opens the back door. What happens in this situation?

14.5 Repeat the analysis of this chapter using a pole of rest length 36 feet and a barn of rest length 60 feet. What special feature is implied by these particular lengths?

CHAPTER FIFTEEN

Voyage to Spica

One fine clear evening last May I set out from my home for a stroll onto campus. At first I admired the brilliant stars of the night sky and the scent of apple blossoms, but after a few minutes I became wrapped up in my own thoughts and oblivious to what was going on around me.

My reverie ended in front of Wilder Hall when I heard muffled sobbing. Despite the darkness, I was able to locate the sound's source. There, sitting on a bench, was my friend and former student Veronica. Her long dark hair, which usually flies free, was pinned up into a tight bun. Her usual energy seemed to have drained away. She looked up at the stars, then buried her face in her hands and gave out another weak little sob. I walked up to her.

"Veronica," I said, "what brings you such despair on this beautiful evening?"

"Oh!" she exclaimed. "I'm sorry. I didn't see you. Oh, please don't worry about me, Mr. Styer, I'm all right." She pulled out a handkerchief and blew her nose loudly.

"No, please, tell me," I countered. "I'd feel terrible if I left you like this when you're so obviously upset. I don't mean to pry, but perhaps if you told me what was bothering you I could do something to help you out."

"Well, it's silly, really, but you're right, maybe you can give me some advice. It all starts with the stars." I was dumbstruck and said nothing. She continued. "Do you see that beautiful blue-white star up there to the south above Dascomb Hall? That's Spica, the brightest star in the constellation Virgo. Don't you see it? Isn't it beautiful? Wouldn't you like to visit there?"

Familiarity with the night sky is not one of my strengths, but eventually Veronica was able to guide my eye to exactly the star she had in mind. I had to admit that it was beautiful, but I also had to admit that I had never considered a personal visit.

"Well, I would love to go there and see it up close for myself," she said. "But I've just looked up its distance in Mudd Library. It's 200 light-years away! I know that I'll never be able to travel faster than the speed of light, c, so no matter how powerful a space vehicle I've purchased, it will take me at least 200 years to travel to Spica. I'll die before I get there!" Her relative composure vanished, and she again dissolved into tears.

"Veronica, Veronica," I tried to comfort her. "It's a good thing you told me your problem, because it's not a problem at all. True, in the Earth's reference frame, it will take at least 200 years for your voyage, and at least 200 years to return. But what's important to you is how much time it takes in *your* reference frame."

"I don't understand," she confessed. "The distance from Earth to Spica is $L_0 = (200 \text{ yr})c$. Because

$$\text{speed} = \frac{\text{distance traveled}}{\text{time elapsed}}$$

and I will travel at some speed V, the time required for my trip will be

$$\text{time required} = \frac{\text{distance traveled}}{\text{speed}} = \frac{L_0}{V} = \frac{(200 \text{ yr})c}{V} = \frac{200 \text{ yr}}{V/c}.$$

To get a small time, I'll need a large V/c, but the largest possible V/c is just less than 1, so the smallest possible transit time is just over 200 years. What could possibly be wrong with my argument?"

"Veronica," I said, "you've correctly calculated the time your trip will require in the Earth's reference frame, using the distance between Earth and Spica in the Earth's frame. But while you travel you'll be in a different reference frame. What's the distance between Earth and Spica in *your* reference frame as you travel?"

"Oh, that's right! I had forgotten. For me, the distance to Spica will be length-contracted!" Veronica's eyes were dry, and the faintest hint of her usual vivacity returned to her face. "So for me, the distance to Spica is

$$L = \sqrt{1 - (V/c)^2} L_0,$$

which can be quite small—indeed, by increasing V, I can make this distance as small as I wish. Right?"

"Yes, that's right." I was pleased to see her perking up. "So now, how much time would it take you to travel to Spica? Not the time in the Earth's frame, but the time in your frame."

"Well, let me see," she puzzled. "I just use the same

$$\text{speed} = \frac{\text{distance traveled}}{\text{time elapsed}}$$

idea that I used before, but now I recognize that, in my frame, Spica is traveling toward me, and it travels over a length-contracted distance. The time required for Spica to reach me is then

$$\text{time required} = \frac{\text{distance traveled}}{\text{speed}} = \frac{L}{V} = \sqrt{1 - (V/c)^2}\frac{L_0}{V}$$
$$= \sqrt{1 - (V/c)^2}\frac{(200 \text{ yr})c}{V} = \sqrt{1 - (V/c)^2}\frac{200 \text{ yr}}{V/c}.$$

How's that?" She smiled in triumph.

"Perfect! Now, suppose you wanted to reach Spica in one year. Could you do it? How fast would you have to travel?" I should learn to keep my mouth shut, but once I start a chain of questioning I find it hard to stop.

"No problem. I just need to set the time I just calculated to 1 year and then solve for V:

$$1 \text{ yr} = \sqrt{1 - (V/c)^2}\frac{200 \text{ yr}}{V/c}$$

$$\frac{1 \text{ yr}}{200 \text{ yr}}(V/c) = \sqrt{1 - (V/c)^2}$$

$$\left(\frac{1 \text{ yr}}{200 \text{ yr}}\right)^2 (V/c)^2 = 1 - (V/c)^2$$

$$\left[1 + \left(\frac{1 \text{ yr}}{200 \text{ yr}}\right)^2\right](V/c)^2 = 1$$

$$(V/c)^2 = \frac{1}{\left[1 + \left(\frac{1 \text{ yr}}{200 \text{ yr}}\right)^2\right]}$$

$$V/c = \frac{1}{\sqrt{\left[1 + \left(\frac{1 \text{ yr}}{200 \text{ yr}}\right)^2\right]}}.$$

That's it!" Veronica's face glowed with excitement. She removed the pin from her hairbun, and waves of luxuriant black hair cascaded freely down her back. "I know I have a calculator in here somewhere." She rummaged through her pack and withdrew a slim silver calculator. It glinted in the starlight as she pressed its buttons. "I can get there in one year, my time, if I travel at speed

$$V/c = 0.999\ 987\ 5."$$

"I'll trust your arithmetic, Veronica. Now, can you reach Spica in just one day?"

"Well, of course. I just use the same formula but with '1/365 year' instead of '1 year.' The answer is

$$V/c = 0.999\ 999\ 999\ 906\ 2.$$

Even faster, but I can do it!"

"What do you mean 'you can do it?' No one can travel that fast!"

"Oh, Mr. Styer, didn't you see?" Veronica said pointing.

In the darkness, I hadn't seen. A few yards from the bench there stood a small space vehicle, elegant and powerful—very much like Veronica herself. I was speechless. Veronica promptly gathered her things, jumped into the cockpit, and made selections from the flight control computer:

Destination? $\longrightarrow$ Spica
Speed (V/c)? $\longrightarrow$ 0.999 999 999 906 2
Seconds to liftoff? $\longrightarrow$ 10

"I'll see you the day after tomorrow!" she sang out from the cockpit. And then she was gone, leaving behind a flash of light and the faint odor of ozone.

No, I thought to myself, you won't see me the day after tomorrow. True, you'll be back here in two of your days, but your voyage will take just over 400 of my years.

I turned to leave and saw a glint from a long straight object on the bench. It was a hairpin. I picked it up and put it in my pocket. The ozone had drifted away, and I could again smell apple blossoms. I walked home admiring the brilliance of the stars.

PROBLEMS

15.1 Veronica calculated the time she would age during her trip using length contraction in her own reference frame. Find the same result by analyzing the trip in the Earth's reference frame and using time dilation.

15.2 Veronica wants to travel to the star Sirius, 8 light-years away, while aging 1 year. How fast must she travel?

15.3 Veronica wants to travel to the star Sirius, 8 light-years away, while aging 8 years. How fast must she travel? Show that travel at this same speed will take her to Spica while she ages 200 years, etc.

15.4 Sketch a graph of the time required for a journey to Spica as a function of the speed V/c using both (a) the commonsense formula L_0/V and (b) the correct relativistic formula derived by Veronica.

15.5 Travel at the speed of light. When I was young, I thought it would be fun to be a flash of light, traveling like a sprite from place to place faster than anything else could travel. Would this really be so exciting? Write a story from the point of view of a burst of light that originates in a flashlight on Earth and is absorbed by an atom within the star Spica. How much time does the journey take? How far must Spica travel in order to reach the bit of light? (You will have to use poetic license, because no reference frame can travel as fast as light does. Instead, approach the limit through a series of reference frames that travel closer and closer to light speed.)

CHAPTER SIXTEEN

Free-for-All

16.1 Relativistic baking. A flat, horizontal tray of cookie dough speeds through a bakery at a substantial fraction of the speed of light. (The dough is square in the tray's reference frame, and thus rectangular in the baker's reference frame.) A baker stands ready with a circular cookie cutter, held horizontally, and as the tray of dough flashes by, she stamps out a cookie with lightning speed. (She drops and raises the cutter so quickly that nothing gets squashed or stuck in the cutter.) The resulting cookie will, in its own reference frame, be shaped like an oval rather than a circle. Is it longer in the direction of its motion or in the direction perpendicular to its motion? Justify your answer in both the tray's frame and the baker's frame. [Source: N. D. Mermin, *Space and Time in Special Relativity* (Prospect Heights, IL: Waveland Press, 1968), p. 229.]

16.2 Relativistic baking: Your own recipe. The previous problem treated the relativistic cookie problem as a "ranking exercise": you were supposed to find out which side of the cookie was longer but not how much longer. Make up a problem that treats this situation but with explicit numbers.

16.3 Spy versus spy (I). James Bond and his archenemy, Goldfinger, each pilot a space ship. Bond flies toward the west, Goldfinger flies toward the east. The two ships are identical except that Goldfinger has mounted a gun at the tail of his ship, pointing perpendicular to the ship, while Bond's ship is unarmed. The two ships fly past each other at high speed.

In an attempt to do away with Bond once and for all, Goldfinger hatches the following diabolical plan: "I'll set my tail gun on a timer to go off at noon, then I'll use my expert piloting skills to guide my ship so that my nose lines up with Bond's tail exactly at noon. My gun will go off and automatically blast the nose off of Bond's ship!"

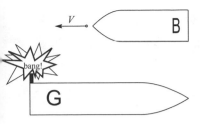

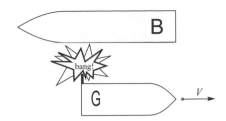

Analysis in Goldfinger's frame   Analysis in Bond's frame

These two sketches suggest that an analysis in Goldfinger's frame indicates that Bond's ship will be safe, while an analysis in Bond's frame indicates that Bond's ship will be hit. Obviously one analysis is wrong. Which is it and why? *Answer in four or fewer sentences.*

16.4 Spy versus spy (II). James Bond and his archenemy, Goldfinger, each pilot a space ship. Bond flies toward the west, Goldfinger flies toward the east. The two ships are identical except that Goldfinger has mounted a gun at the tail of his ship, pointing perpendicular to the ship, while Bond's ship is unarmed. Each ship is 150 feet long in its own rest frame. The two ships fly past each other with a relative speed of $V = \frac{4}{5}c$.

In an attempt to do away with Bond once and for all, Goldfinger hatches the following diabolical plan: "I'll set my tail gun on a timer to go off at noon, then I'll use my expert piloting skills to guide my ship so that my nose lines up with Bond's tail exactly at noon. My gun will go off and automatically blast the nose off of Bond's ship!"

Bond hears about this plan from a friend, the lithe and lovely Christine, planted in Goldfinger's crew. But it doesn't worry him: "My ship will be length-contracted, and Goldfinger's bullet will fly harmlessly in front of my bow," Bond claims. (Goldfinger is a man of limited intelligence who has never heard of length contraction.)

   a. In Goldfinger's frame, how long is Bond's ship? Sketch the situation in Goldfinger's frame.

   b. In Goldfinger's frame, how far in front of Bond's bow does the bullet fly?

Christine is more concerned about Bond's safety than Bond himself is. "But in *your* frame, Goldfinger's ship is contracted. Doesn't that mean that you'll be hit?"

   c. In Bond's frame, which happens first: the firing of the bullet or the nose-tail lineup? Draw *two* sketches in Bond's frame that correspond to the one sketch from part (b).

   d. Show that in Bond's frame, the timer at the tail of Goldfinger's ship is set 120 nans ahead of a clock in the nose of Goldfinger's ship.

   e. While Goldfinger's clocks tick off 120 nans, how much time elapses in Bond's frame?

    f. How far does Goldfinger's ship move in Bond's frame while Goldfinger's clocks tick off 120 nans?

    g. In Bond's frame, the bullet again flies harmlessly in front of Bond's bow. How far in front of the bow does it pass?

    h. How can we understand the difference between the results of parts (b) and (g)? Suppose Bond's ship had been outfitted with a 180-foot flagpole protruding straight out from the bow. Goldfinger's bullet, although still nonfatal, would in this case snap the flagpole. Part (g) gives the length of the remnant in Bond's frame, while part (b) gives the length of the remnant in Goldfinger's frame. Are these two lengths related as you would expect?

16.5 Spy versus spy (III). Consider the same situation as in problem 16.4, but suppose that each ship has rest length L_0 and that they approach each other with relative speed V. How far does the bullet pass in front of Bond's bow in Bond's frame? In Goldfinger's frame? Do these expressions have the expected relation for all values of V and L_0? Do they have the expected behavior when $V \ll c$?

16.6 Train in a tunnel (I). A train is 900 feet long in its own frame, and a tunnel is 1500 feet long in its own frame. Ann sits at the very front of the train, and Rodger sits at the very rear. The train speeds from west to east through the tunnel at $V = \frac{4}{5}c$.

    In the train's frame the moving tunnel is length-contracted to be exactly 900 feet long [900 ft $= \frac{3}{5}(1500$ ft$)$]. Ann and Rodger both stick their heads out the window and glance up at the instant the train exactly fits within the tunnel, so Ann sees the east portal of the tunnel and Rodger sees the west portal of the tunnel.

    These questions deal with the situation in the tunnel's frame.

    a. In the tunnel's frame, how long is the train?

    b. In the tunnel's frame, Rodger glances before Ann. By how much is Rodger's clock set ahead of Ann's?

    c. While Rodger's watch ticks off the time determined in part (b), how much time elapses in the tunnel's frame? (This is the time elapsing between Rodger's glance and Ann's glance.)

    d. During the time between glances, how far does the train move (in the tunnel's frame)?

    e. Sum your answer to part (a) and your answer to part (d) to find Ann's position when she glances up. How does this compare with the situation as found in the train's frame?

16.7 Train in a tunnel (II). Carry out the previous problem algebraically using a tunnel of rest length L_0 and a train of rest length $\sqrt{1 - (V/c)^2}\,L_0$ moving at speed V.

16.8 Busted bus. A bus 30 feet long in its own reference frame moves so fast that it is only 6 feet long in the Earth's reference frame. It speeds through a "falling rock zone," and as bad luck would have it, a rock (15 feet long in its own frame) rolls down the mountainside and completely crushes the 6-foot-long bus.

Or does it? In the reference frame of the bus, the rock is 3 feet long. Does the rock punch a neat 3-foot hole in the roof of the 30-foot bus?

This is not a matter about which we can say "both are right": the bus is either completely squashed, or else it is damaged but not destroyed. Which analysis is correct?

16.9 Falling hula hoop. A horizontal hula hoop, one yard wide in its own frame, falls slowly from the ceiling.* As it falls, a horizontal yardstick is launched across the room so fast that it is only 1 foot long in the room's frame. The launch is timed so that the hula hoop momentarily surrounds the shrunken yardstick. The length-contracted yardstick easily passes through the center of the hula hoop.

What happens in the yardstick's frame? In this frame the hula hoop is length-contracted. How can a 3-foot-long yardstick pass through the center of a 1-foot-wide hula hoop?

16.10 Explosion! I have a U-shaped piece of hardened steel, and I place high explosive into the hole. I also have a T-shaped piece of the same hardened steel, and on the bottom tip of the T, I mount a detonator. The T is just short enough that I can gently slide it into the U, and the detonator will not touch the explosive.

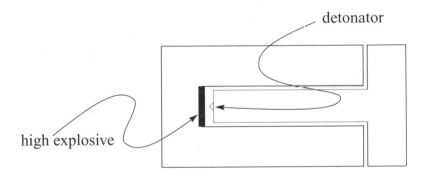

Now I hurl the U and the T toward each other at a substantial speed. In the reference frame of the T, the U is length-contracted, so when the two pieces come together the detonator and explosive touch. The U and the T are both destroyed in the resulting explosion.

---

*For non-Americans: A yard is about a meter.

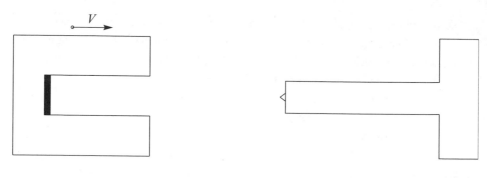

But wait! In the reference frame of the U, the T is length-contracted, the detonator does not touch the explosive, and both pieces remain intact.

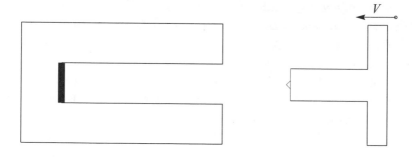

Clearly, these two analyses cannot both be correct. Find the flaw (a hidden assumption) in one analysis, and state definitively whether the explosion goes off or not. [Source: E. F. Taylor and J. A. Wheeler, *Spacetime Physics*, 2nd ed. (New York: W. H. Freeman, 1992), p. 185.]

16.11 Stealth tank. A tank is an important military vehicle for two reasons: first, it carries a gun; and second, it can travel over rough terrain dotted with holes and ditches.

A military contractor approaches a Pentagon procurement officer with a proposal to build a "stealth tank" that can travel at relativistic speeds. The contractor points out all the advantages of the tank: its armament, its speed, its evasiveness. Then he comes to the clinching point of his sales pitch: "Officer, this tank is 15 feet long in its own frame, so if it were an ordinary tank attempting to go over a 15-foot-wide hole, it would fall in. But it moves so fast that in its reference frame that the hole is length-contracted to be only 3 feet wide. Our tank will roll right over that little hole!"

The procurement officer has heard similar extravagant claims from other contractors, so she thinks about it for a while. "Wait a minute," she says, "in the reference frame of the hole, the hole is 15 feet wide and the tank is length-contracted to be only 3 feet long. I'm sure that your stubby little tank will be trapped in such a wide hole."

Who's right, the contractor or the procurement officer?

16.12 Fatal mistake (II). Consider again the situation in problem 14.2, "Fatal mistake." Suppose the pole is as rigid as possible, so that the compression within the pole travels at the speed of light. In the pole's frame, how far is the pole's butt from the front door when it gets the message that the back door didn't open?

16.13 Joust. Sir Lancelot and Sir Mordred each has a lance of rest length 17 feet. During a joust, each gallops toward the other at speed $\frac{3}{5}c$ relative to the Earth. (For definiteness, assume that Lancelot gallops to the right, $v_L = +\frac{3}{5}c$, while Mordred gallops to the left, $v_M = -\frac{3}{5}c$.)

    a. The crowd in the stands sees a fair fight because each knight has a lance of equal length. How long is each lance in the Earth's frame?

    b. How fast is Mordred moving in Lancelot's frame? How long is Mordred's lance in Lancelot's frame?

    c. How fast is Lancelot moving in Mordred's frame? How long is Lancelot's lance in Mordred's frame?

In Lancelot's frame, his own lance is much longer than his opponent's, so Lancelot expects to win easily. However, the same holds for Mordred! Meanwhile, in the crowd's frame, the two lances are equally long.

In the crowd's frame, each lance makes contact with the opposing knight's breastplate simultaneously. Also, both knights are unhorsed simultaneously.

    d. Which knight is unhorsed first in Lancelot's frame? In Mordred's?

    e. Is Lancelot unhorsed at the instant that his breastplate makes contact with Mordred's lance?

    f. In each knight's frame, one knight is unhorsed by another knight who has already been struck. How can a struck knight succeed in unhorsing his opponent?

[Inspired by Paul Horwitz, Edwin Taylor, and Kerry Shetline, *RelLab User's Manual* (New York: Physics Academic Software, 1993), pp. 68–69.]

16.14 The paradox of the mirror. This book started by considering the question, "If I were in a railroad coach moving at the speed of light, and I looked into a mirror, then what would I see?" Resolve this paradox. (Clue: See problem 15.5.)

PART IV / Starting and Stopping

CHAPTER SEVENTEEN

General Relativity

Albert Einstein produced his theory of special relativity in 1905. A good theory not only answers old questions but also generates new ones, and by this measure special relativity proved to be a very good theory indeed, because it immediately generated a flood of significant questions for scientists to work on. Two of them were as follows.

First, a question likely to occur to anyone: We got a lot of mileage through the principle of relativity relating observations in two different inertial reference frames. But what about noninertial ("accelerated") frames? It's clear that identical internal experiments performed in two frames, one inertial and one accelerating, will *not* give identical results. For example, if you pour coffee in an inertial frame, it falls into the cup below, but if you pour coffee in an accelerating frame, it ends up on your shirt. But is there some different principle that *does* apply to noninertial frames?

Second, a question likely to occur to any physicist: In Newton's theory of gravity, the force of attraction between two distant bodies (say, the Sun and the Earth) depends on the mass of each body (greater mass results in more attraction) and on the distance between them (greater distance results in less attraction). Nothing else matters. If the Sun were suddenly to move closer to the Earth, then the gravitational force on the Earth would increase, and (because "nothing else matters") that force would increase *at the same instant* that the Sun moved. This is a way to send a message instantaneously, but as we saw in chapter 10, "Speed Limits," no message can be sent instantaneously. Obviously Newton's theory, while an excellent approximation, is inconsistent with special relativity and requires modification.

These two questions appear to me to be utterly unrelated. I confess that if I had encountered them in 1905, I would have thrown up my hands in despair and turned my mind to something less confusing. But as we've already seen several times in this book, Einstein had both more insight and more courage than I do. Einstein explored the possibility that the two questions might have something in common.

If I sit in a jetliner that accelerates down the runway, and pour coffee, that coffee ends up on my shirt. Einstein asked: Is there any other way that poured coffee could end up on my shirt? Suppose that the jetliner were stationary on the runway, but a large mass suddenly materialized on the runway behind the jetliner: say, a big asteroid were gently placed there. Then the gravitational attraction between the asteroid and the coffee would make the coffee fall, not straight toward the center of the Earth, but somewhere between straight down and straight back. This is another way that would make the poured coffee end up on my shirt.

As Einstein thought through these parallel situations, he couldn't find any way (other than looking out the window) to tell the difference between a jetliner accelerating down a runway and a jetliner stationary on a runway with a large gravitating mass behind it. He hit upon the *principle of equivalence:*

> *A reference frame stationary in gravity and one far from any matter but accelerating at an appropriate rate will give rise to identical results for all internal experiments.*

Consequences of the Equivalence Principle

Here's one direct consequence of the equivalence principle. Suppose you're in a rocket ship far from any planet or star. Your rocket ship's engines are blasting, so you're accelerating. You measure your speed relative to a distant star and find that it's 530 miles/hour. A second later, it's 550 miles/hour. A second later still, it's 570 miles/hour. And so forth. We say the rocket ship accelerates at 20 (miles/hour)/second.

Now, suppose that you stand up tall, stretch out your arm, and when the ship's speed reaches 530 miles/hour, release a house key. The key continues to move upward at 530 miles/hour. But soon the spaceship floor, and your arm, are no longer moving at 530 miles/hour. Just one second after release, the spaceship floor is moving upward at 550 miles/hour. The floor rises up toward the key. Just one second after release the key is moving downward relative to the floor at 20 miles/hour. Two seconds after release, the key is moving downward relative to the floor at 40 miles/hour. Sooner or later, the key and the floor meet: you can say either that the floor rises to meet the key or that the key falls to hit the floor.

We've shown that if a key is released in "a reference frame far from any matter but accelerating," the key will hit the floor. According to the principle of equivalence, this means that if a key is released in "a reference frame stationary in gravity," that key will likewise hit the floor. Sure enough, this prediction is correct—but it is not particularly surprising or enlightening.

However, a small twist on the experiment makes it quite interesting. Suppose that in your rocket ship you release a bowling ball instead of a key. Once again, and in exactly the same way, the floor rises to meet the bowling ball, as shown in figure 17.1. The interesting part is the phrase "in exactly the same way." The floor will rise up to meet the bowling ball in exactly the same amount of time that it previously took to meet the key. In fact, it doesn't matter whether you release a key or a bowling ball or a jar of marmalade: all these objects will meet the floor in the same amount of time. This is because it's an issue of the floor rising, and of course the floor rises in just the same way regardless of whether you have released a key or a bowling ball or a jar of marmalade.

According to the principle of equivalence, the same result must apply for keys, bowling balls, and marmalade jars dropped in "a reference frame stationary in gravity" such as on the Earth's surface (fig. 17.2). If the equivalence principle is correct, all will hit the floor after the same amount of time.

Now this result is unexpected. It's quite natural to think that a heavy bowling ball will hit the floor in less time than a light key, but if the equivalence principle is correct, then the ball will not hit the floor earlier.

According to legend, Galileo dropped from the Tower of Pisa a musket ball weighing half an ounce and cannonball weighing 100 pounds and found that they struck the ground at the same time. From a modern perspective, this was a test of the equivalence principle. Nowadays you can test the principle of equivalence yourself in dramatic experiments such as the Tower of Terror at Disney World. In these experiments members of the thrill-seeking public are dropped, and they can determine for themselves that their big and small friends, as well as their keys and change and cell phones, all hit the ground at the same time.

The equivalence principle has been tested numerous times since Einstein advanced it in 1907. In 1986, physicist Ephraim Fischbach and three colleagues suggested that under certain circumstances it would fail, and this suggestion resulted in a whole battery of new tests. So far, the equivalence principle has yet to fail. Physicists at the University of Washington are continuing a series of high-precision

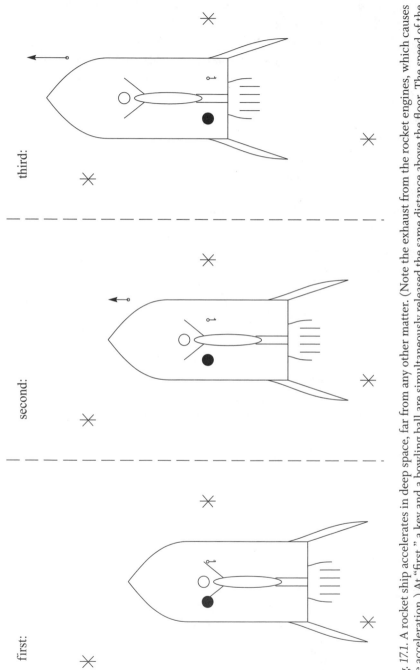

first:

second:

third:

Fig. 17.1. A rocket ship accelerates in deep space, far from any other matter. (Note the exhaust from the rocket engines, which causes the acceleration.) At "first," a key and a bowling ball are simultaneously released the same distance above the floor. The speed of the rocket ship relative to the stars at this instant happens to be zero. At "second," the key and bowling ball are still motionless with respect to the stars, but the floor of the rocket ship is rising to meet them. At "third," the key and bowling ball are still motionless, and the floor is rising even faster.

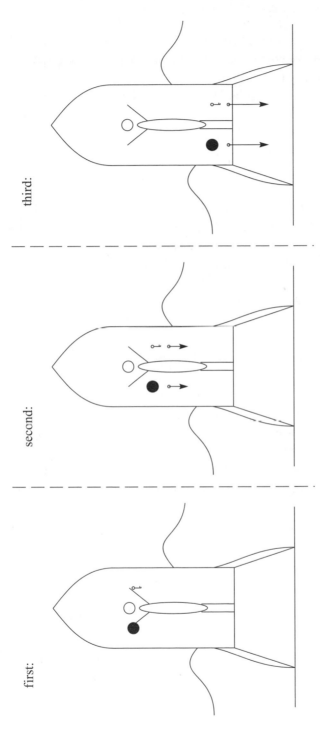

first:

second:

third:

Fig. 17.2. A rocket ship is stationary on the surface of the Earth. (Note the absence of rocket exhaust.) According to the equivalence principle, the key and bowling ball will hit the floor simultaneously, just as they did in the case of acceleration in deep space.

experimental tests. Someday, perhaps, a situation will be discovered in which the equivalence principle doesn't work. This would be an exciting day, because then scientists would have to figure out what principle should replace the principle of equivalence. But even if an exception is found, the equivalence principle will remain a useful idea that holds approximately in a wide variety of situations, just as flat maps are useful even though the Earth is round.

The Bending of Light by Gravity

Figures 17.1 and 17.2 show the simultaneous release of a bowling ball and a key. The same experiment could, of course, be performed with a bowling ball and a jar of marmalade. What happens if it's performed with a bowling ball and a bit of light? We cannot, of course, hold a puddle of light in our hands and then release it. But we could release the bowling ball and simultaneously flick a horizontal flashlight on and then off to create a short burst of light.

It's clear what would happen in deep space: the floor of the rocket ship would rise to meet the light, just as it rises to meet the bowling ball, the key, or the marmalade. (Because the light is traveling horizontally at the same time the floor is rising vertically, the light might hit the side wall before the floor has a chance to rise up enough. To solve this problem we can just procure a wider rocket ship. While this would be expensive in practice, these thought experiments are free, so budget is not a concern.) According to the principle of equivalence, the same result must apply to a light burst launched horizontally on the Earth's surface.

We are used to thinking that light travels in a straight line. But if the foregoing analysis is correct, then a light ray bends toward a gravitating mass. The curvature isn't dramatic: on the surface of Earth, in one second light travels horizontally by 186 000 miles and drops vertically by 16 feet (which is exactly how much a bowling ball drops in one second). This is so small that the curvature of light rays is not measurable in most circumstances.

Is there a way to test his spectacular and unexpected conclusion? To maximize light deflection, the light should pass near a very massive object, and the most massive object nearby is the Sun. The light ray from a star positioned behind the Sun and near its edge would be bent more than any other light ray we're likely to encounter.

How, though, could anyone see the light from such a star? During the day, when the Sun is visible, the stars are not. There's only one exception to this rule, and that's during a solar eclipse.

In 1911 Einstein predicted that that path of light would bend under the influence of gravity. Astronomers interested in testing his prediction during a solar eclipse were stymied by the outbreak of World War I in 1914. In 1916, after more thought, Einstein realized that he had made an error and that light would bend twice as much as he had predicted in 1911. The war ended by armistice in 1918, paving the way for an observational test during the eclipse of 29 May 1919.

The story of that test is a convoluted and fascinating one, well told by the expedition organizer Sir Arthur Eddington (see the references). But the scientific result is straightforward: the light did indeed bend. The amount of bending was minute and thus difficult to measure, but within the limits of observational error the measured bending matched Einstein's prediction of 1916.

The deflection of light by the Sun was just barely detectable using the technology of 1919; with today's technology it can be measured easily. In 1970, for example, scientists measured both the bending and the timing of radio signals from the *Mariner 6* space probe as it passed behind the Sun. The results were in accord with Einstein's theory.

More recently, objects more massive than the Sun have been recruited to study the bending of light. Abell 2218 is a cluster of galaxies located about three billion light-years from Earth. It is about a trillion times more massive than the Sun. It just so happens that directly behind Abell 2218 is another galaxy. Some light from that background galaxy heads out above the cluster, but gravity from the cluster bends it down to strike Earth. Other light heads out below the cluster, but gravity from the cluster bends it up to strike Earth. Similarly for light heading out to the right and left of the cluster. The photograph in figure 17.3 shows Abell 2218 surrounded by several long faint arcs—the light from a single background galaxy that has been bent by gravity like a lens.* No longer a tiny effect, in this photograph the bending of light by gravity is readily apparent to even a casual glance.

---

*If Abell 2218 were perfectly spherical and if the background galaxy were dead-center behind it, then the background galaxy would appear as a perfectly circular "Einstein ring" centered on Abell 2218. Because these conditions do not exactly hold, the background galaxy instead appears as several arcs.

Fig. 17.3. The Abell 2218 cluster bends the light of a galaxy behind it, stretching that galaxy's image into several arcs. This phenomenon in called "gravitational lensing." Courtesy of NASA, European Space Agency, Andrew Fruchter and the Early Release Observation team (Space Telescope Science Institute, European Coordinating Facility for the Space Telescope).

Gravitational Time Dilation

Einstein proposed another thought experiment to be performed in the accelerating deep space frame, and this thought experiment unveiled another surprising fact concerning the behavior of time.

Suppose the rocket ship in deep space is outfitted with two clocks, one near the tail and one near the nose (fig. 17.4). The tail clock is programmed to release two very brief light bursts, one at the beginning and one at the end of a nan. The nose clock is programmed to receive these two light bursts and to note the amount of time it ticked off between these two receptions. We analyze this experiment in the inertial frame that moves at the same speed as the rocket ship when the two bursts were released.

The two light bursts are 1 foot apart as they travel up toward the nose. The nose clock receives the first burst. But by now, the nose clock is moving (in the inertial frame of our analysis); it is running away from the second burst. After the first burst is received, the second burst has to travel more than 1 foot before it reaches that withdrawing nose clock. Therefore, the nose clock has to tick off more than 1 nan between receiving the two light bursts.

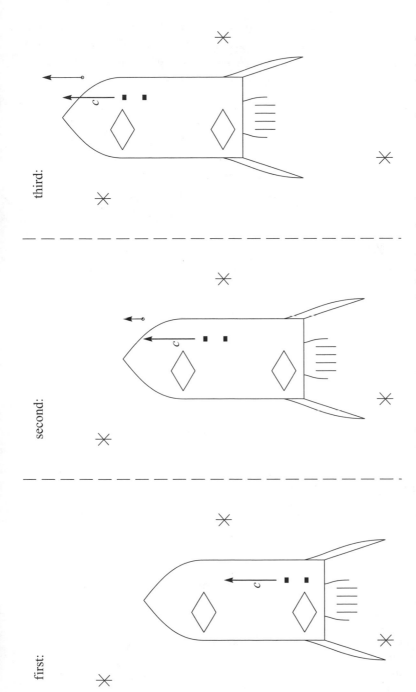

Fig. 17.4. A rocket ship accelerates in deep space, far from any other matter. A clock in the tail sends a timing signal to a clock in the nose. All three diagrams are drawn in the inertial reference frame that moves along with rocket ship when the tail clock emits the two signals.

In a rocket ship accelerating in deep space, the nose clock must tick off more than 1 nan while the tail clock ticks off exactly 1 nan. According to the principle of equivalence, the same result must be obtained by a pair of stationary clocks stacked in gravity. This result is called *gravitational time dilation:*

A lower clock ticks more slowly than a higher clock.

In this book, we have not worked out exactly how much more slowly the lower clock ticks, although you can surmise that the difference is minute under ordinary circumstances. Nevertheless, it has been detected using clocks of extreme accuracy. For example, a 1960 experiment by R. V. Pound and G. A. Rebka confirmed gravitational time dilation by comparing two natural clocks—the internal vibrations of iron nuclei—one in the basement and one on the top floor of Harvard University's Jefferson Physical Laboratory.

We have already seen that the Global Positioning System needs to be cognizant of motional time dilation (see page 71). It is equally true that the GPS needs to be cognizant of gravitational time dilation, and indeed this awareness is fully integrated into the system. Without that awareness, the system would fail completely.

REFERENCES

Experiments testing general relativity are described in

Clifford M. Will, *Was Einstein Right? Putting General Relativity to the Test*, 2nd ed. (New York: Basic Books, 1993).

The dramatic story of Eddington's 1919 eclipse expedition to see whether light is deflected by the Sun's gravity is told in

Sir Arthur Eddington, *Space, Time and Gravitation* (Cambridge: Cambridge University Press, 1920), chap. 7.
Subrahmanyan Chandrasekhar, *Eddington: The Most Distinguished Astrophysicist of His Time* (Cambridge: Cambridge University Press, 1983).

In recent years, a number of misconceptions concerning the analysis of the 1919 eclipse data have arisen within science history circles. Most of these misconceptions are put to rest in

Matthew Stanley, "'An expedition to heal the wounds of war': The 1919 eclipse and Eddington as Quaker adventurer," *Isis*, 94 (2003) 57–89.
Daniel Kennefick, "Testing relativity from the 1919 eclipse: A question of bias," *Physics Today*, 62 (March 2009) 37–42.

CHAPTER EIGHTEEN

A Pair of Clocks Starts Moving

We return to the question asked on page 45: If Veronica starts out at rest in Ivan's frame with two clocks that are synchronized, how does it happen that when she speeds up, the clocks remain synchronized in her frame but fall out of synchronization in Ivan's frame?

Suppose that Veronica has two clocks, and she accelerates to the right with them. In Ivan's frame, both clocks accelerate at the same rate for the same amount of time, so at any instant both are moving at the same speed, and both tick slowly by the same amount. If nothing else were done, they would remain synchronized. In Veronica's frame, the acceleration to the right is equivalent to gravity pulling left. Thus, the clock lower down in gravity—the clock on the left—ticks slowly relative to the one on the right. If nothing were done, Veronica's two clocks would fall out of synchronization. To prevent this, during the acceleration process Veronica must manually adjust her left clock, turning its time forward in order to keep her two clocks synchronized. Ivan's frame is not accelerating, so he doesn't see the gravitational effect; he just sees Veronica turning her left clock forward. Because of Veronica's manual adjustments, her left clock—the rear clock in Ivan's frame—is set ahead of her right clock in Ivan's frame.

Ivan's frame

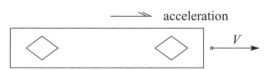

Veronica's frame

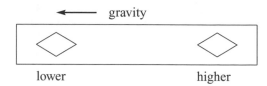

We discussed the twin paradox in chapter 13. (See figs. 13.1 and 13.2) Veronica travels to Sirius and back while Ivan remains at home on Earth. From Ivan's perspective, Veronica travels a distance of 8 light-years to Sirius at speed $V = \frac{4}{5}c$. This journey requires 10 years, but because her clock ticks slowly, Veronica ages only 6 years. Veronica then takes 2 weeks to turn around and returns to Earth at the same speed. When she returns, she has aged 12 years while Ivan has aged 20 years. Ivan explains this by saying that Veronica's clock ticks slowly.

From Veronica's perspective, Sirius travels toward her over the length-contracted distance of 4.8 light-years. This journey requires 6 years, during which time the moving Ivan ages only 3.6 years. Similarly for the return journey. But during her 2-week turnaround, the Earth clock changes from being the front clock, set *behind* the Sirius clock by 6.4 years, to being the rear clock, set *ahead of* the Sirius clock by 6.4 years. In other words, during Veronica's 2-week turnaround, Ivan ages 12.8 years. The total number of years ticked off by the Earth clock through this whole process is 3.6 + 12.8 + 3.6 = 20.

The two points of view obtain identical results, as they must. But they explain these results in different ways: From Ivan's point of view Veronica's clock ticks slowly. From Veronica's point of view Ivan's clock usually ticks slowly, but it jumps ahead rapidly during the turnaround.

This resolution of the paradox is correct, but it lacks insight. How, to Veronica, can Ivan's clock tick off 12.8 years during the turnaround while her clock ticks off only 2 weeks? The answer lies in gravitational time dilation. Look at figure 13.1. When Veronica is slowing down near Sirius, she will be tossed forward against her seat belt. When she is speeding up to begin her homebound journey, she will be tossed backward into the back cushion of her seat. Both of these effects are equivalent to gravity pushing to the right. Thus, during the turnaround, Veronica's clock is a lower clock, and through gravitational time dilation, it ticks slower than the high Earth clock.

REFERENCES

The assertions made in this chapter are backed up by quantitative, analytic arguments in

Daniel F. Styer, "How do two moving clocks fall out of sync? A tale of trucks, threads, and twins," *American Journal of Physics*, 75 (2007) 805–14.

CHAPTER NINETEEN

Black Holes

Y ou have probably noticed that the two previous chapters contained no equations: I said that a lower clock ticks slowly but didn't say how slowly. I said that the path of a light beam bends under the influence of gravity but didn't say by how much. These equations have been worked out, but it's beyond the scope of this book to work them out here.* Consequently, we don't have the tools on hand to answer some interesting questions like How low can you go? before your clock slows down to a specified rate. I can't resist, however, telling you some of the surprising results of general relativity. So in contrast to the rest of this book, this chapter is descriptive rather than analytic. I'll just tell you the results of calculations that have been performed, rather than show you the experiments, observations, and reasoning that underlie those results.

A black hole is an entity so dense that nothing from inside the "hole" can ever exit—not even light. Black holes have several properties: mass, electric charge, and spin. Here, we'll consider only the simplest case, namely a spherically symmetric black hole with no electric charge and no spin.

Some of the properties of a black hole follow simply from its powerful gravitational attraction. It is surrounded by a spherical surface called the "Schwarzschild surface" or the "event horizon," through which nothing from inside can exit.† A rocket

---

*Although if you invest a moment's thought, you'll realize that gravitational time dilation—a change of time with height—must be accompanied by a change of length with height.

†Karl Schwarzschild uncovered the relevant solutions to Einstein's equations of general relativity in December 1915, just 1 month after Einstein had adduced the equations. This feat is all the more astounding in that Schwarzschild was at the time (during World War I) serving in the German army as an artillery lieutenant on the Russian front. One can speculate that Schwarzschild immersed himself in this scientific work in order to assuage the horrors of war.

launched inside can never get outside because even the most powerful rocket will be pulled back down toward the center. A tower erected on the inside cannot be built up as tall as the event horizon because even the strongest building materials will be flattened by the gravitational attraction. A light flash originating inside the event horizon cannot get outside because the gravity draws even light flashes to curve back down.

The size of this surface depends only on the black hole's mass: the more massive the hole, the larger the surface.

Orbiting a Black Hole

Ivan, Veronica, and Veronica's sister Denise decide to study a particular black hole, a highly massive one. They purchase a large green space ship that has a small green shuttle stored within the ship's shuttle bay. Being careful to stay well outside the event horizon, the trio pilot their space ship into orbit around the black hole.*

This expedition, while expensive, is perfectly safe. There's a common misconception that a black hole somehow reaches out and sucks in unsuspecting spacecraft, but in fact the gravity far from a black hole is identical to the gravity from an ordinary body like the Sun. As long as our space travelers are far away, staying in orbit around a black hole is no more taxing than staying in orbit around the Sun.

In order to understand this orbit around a black hole, we must first answer two questions about traditional orbits around the Earth, namely, Why don't satellites fall down? and Why are astronauts weightless? We begin by thinking about how objects get into orbit.

How can I put a baseball into orbit around the Earth? I begin by standing outdoors and dropping a baseball. It falls and lands at my feet—this is not in orbit. So I try again, but this time I pitch the baseball horizontally. Now it both falls and moves over, and this time it lands 20 yards away from me—again it's not in orbit. I try a third time, with a faster horizontal pitch. Again the ball both falls down and moves over, but this time it lands 500 yards away from me.† I try a fourth time, with a *very* fast horizontal pitch. Again the ball both falls down and moves over, but this time

---

*They select an orbit far enough from the black hole that their orbital speed (with respect to the black hole) is much less than the speed of light. That way they don't need to worry about special relativistic time dilation or length contraction or any other motional relativistic effect. This allows them to concentrate solely on gravitational relativistic effects.

†To all professional baseball talent scouts: I'm not really this strong a pitcher. These are thought experiments.

it's moving so fast that it falls over the horizon. And what's on the other side of the horizon? Another horizon! The ball is falling down, but the Earth is curving down as well. This last pitched ball keeps falling at the same rate that the Earth keeps curving. This last pitched ball is in orbit around the Earth.

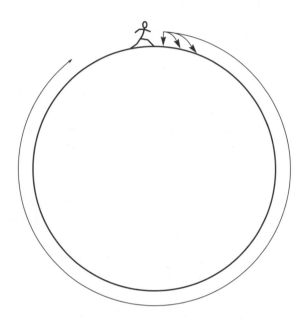

The previous paragraph, while correct in principle, ignores two practical considerations: First, the baseball-in-orbit scenario works when, after the ball is pitched, the only force acting on it is gravity. It would not work if the baseball slammed into a tree. Nor would it work if the baseball encountered a lot of air. A baseball could orbit the Moon skimming just above the Moon's rock surface, but a baseball orbiting the Earth must orbit above the Earth's atmosphere. Second, not even the greatest professional baseball pitcher can pitch a ball at the speed needed to get it into orbit, which happens to be (at the Earth's surface) 17 700 miles/hour. That's why we need rockets: both to lift the satellite above the atmosphere and to give it a horizontal pitch sufficient to keep it falling over the horizon time and time again.

This analysis lets us answer our two questions. First, why don't satellites fall down? They *do* fall down, but they move over horizontally at the same time, so they fall down over the horizon. Once the satellite is pitched into an orbit above the atmosphere, the only force acting upon it is gravity: it no longer fires its rocket engines.

Second, why are astronauts weightless? It's not because there's no gravitational force acting on them.* When an astronaut is in the International Space Station, his gravitational attraction toward the Earth is only 10% less than it was on the Earth's surface. The astronaut falls toward the Earth, but the space station itself falls toward the Earth at this same rate. The keys, bowling balls, marmalade jars, and citizens in the space station all fall toward the Earth at this same rate. If someone were to open a space station window and release a wrench, the wrench would fall toward the Earth at this same rate. Thus, the wrench would hover right outside the window: it would fall relative to the Earth, but it would not fall relative to the space station, because the wrench and the space station are falling at the same rate relative to the Earth. It is in this sense that astronauts (and bowling balls and wrenches) in the space station are "weightless."

Returning to our trio in orbit around the black hole, we can understand that they are "weightless" not in the sense that they don't fall toward the black hole, but in the sense that all three explorers fall toward the black hole at the same rate that their space ship falls toward the black hole. If they shuttered their windows, they couldn't tell whether they were orbiting a black hole or orbiting the Sun or at rest far from any matter.

But when they unshutter their windows, they do realize the difference. When they peer toward the center of their orbit they see, not the bright warm Sun, but an inky blackness. Careful measurements of the positions of stars nearly behind the black hole show that the light from these stars does not travel in straight lines; instead, the light is bent toward the black hole as it passes nearby. Occasionally the black hole passes directly between the space ship and a distant star, and when that happens the distant star is visible as an Einstein ring around the black hole.

Controlled Descent to Near the Event Horizon

At first, our explorers are content to orbit the black hole and rejoice in the novelty of orbiting a body that neither generates nor reflects light, but only absorbs it. Denise, however, grows curious to approach the black hole more closely.

---

*For example, the International Space Station orbits at an altitude of 220 miles, which is high by jetliner standards but only 6% of the radius of the earth. The downward acceleration of a falling ball at the Earth's surface is 9.8 meters/second^2 = 22 mph/second—that is, a ball dropped from rest travels at 22 mph after 1 second, 44 mph after 2 seconds, 66 mph after 3 seconds, and so forth. The downward acceleration of the International Space Station is not much less: 8.8 meters/second^2 = 20 mph/second.

Leaving Ivan and Veronica in the mother ship, she departs in the shuttle. Her first act is to fire the shuttle rockets into the direction of her orbital motion, until she comes to a complete stop: she is at rest relative to the black hole. Now she starts to fall straight down into the black hole, but she aims her rockets downward and adjusts their thrust to exactly balance the attraction of gravity, so she remains at rest relative to the black hole. She feels an unfamiliar sensation: weight! Because she and her spacecraft are no longer falling together, she is no longer weightless.

She turns her rocket thrust down just a bit, and slowly drops toward the black hole. She keeps her rate of descent very small, and whenever she wishes, she just turns up the rocket thrust to stop completely. In this way she makes a controlled descent toward the black hole.

What does Denise see? What does she feel? First of all, she feels heavier and heavier as she approaches the black hole. She notices also that she needs more and more rocket thrust to keep her rate of descent small. As she approaches the black hole, the inky, starless blackness fills up more and more of the sky beneath her feet. The increasing gravitational attraction bends light from distant stars more and more. Familiar constellations in the direction of the black hole appear at first vaguely disconcerting, then distorted, and then totally unrecognizable. An ever greater portion of the sky is distorted in this way. At one point, she looks out horizontally and notices a small green shuttle in the sky. What? Is some other explorer descending toward this same black hole? No. The shuttle she sees has the same registration number as her own. She is seeing the light from her own shuttle, traveling in orbit around the far side of the black hole and coming right back to her eye.

As Denise descends, she occasionally looks up toward Ivan and Veronica in the mother ship. Each time it completes an orbit, the mother ship appears directly above Denise's head. Denise notices that that these reappearances are coming more and more rapidly: her low clock is ticking slowly, so as she descends less and less time ticks off on Denise's clock between these occurrences, which are, to Ivan and Veronica, always separated by the same amount of time.

At one point, Denise looks up, looks again, then rubs her eyes and looks a third time. The mother ship appears not green, but blue! To Denise, all the clocks on the mother ship tick quickly, and that includes the clocks that are atoms radiating green light. Blue light has a smaller period than green, so to Denise these atoms are radiating blue light. As Denise descends, the mother ship turns from green to blue to violet. As she descends still further she can no longer see the mother ship with her eyes, because its light is ultraviolet. Fortunately, she brought along ultraviolet detecting goggles, and using them she can still track the motion of the mother ship.

Denise continues her controlled descent. Her weight is difficult to bear. The shuttle rocket engines are thrusting mightily. She turns up the rocket thrust just an iota to halt her descent and rest stationary immediately above the event horizon. Everything below her feet is black. She shines a flashlight down into the blackness, and nothing reflects back. She attempts to snag something from under the event horizon by dangling a fishing line, but the fishing line snaps under the intense gravitational pull. She tries stronger fishing wire, and that snaps too. She wisely declines to dip her arm into the blackness.

And with that decision, she increases the rocket thrust and slowly makes her way back up to the mother ship. The mother ship turns from ultraviolet to violet to blue to green again. When she reunites with Ivan and Veronica, they both remark on how young Denise is, because they have aged more than she has.

Swapping Stories

After Denise tells her story to Ivan and Veronica, she asks them to fill her in on their observations. "Mostly it's been just boring, orbiting this hole time and again awaiting your return. When we watched you, you were going about your tasks more and more slowly: your heart was beating slowly, the hands of your watch were rotating slowly. Your shuttle turned from green to orange to red to infrared, and we had to follow your progress watching through our infrared detecting goggles."

Falling into a Black Hole

Ivan is mesmerized by Denise's report. If black holes are this unusual on the outside, what are they like on the inside? Deciding that he must find out, he develops a plan. Veronica and Denise, knowing that they will never see Ivan again, tearfully agree. Veronica remains up in orbit in the mother ship. Denise once again boards the shuttle and makes a controlled descent until she rests just outside the event horizon. Then Ivan enters the ship's green escape pod, detaches from the mother ship, and blasts the pod's rockets into the direction of his orbital motion until he comes to a complete stop with respect to the black hole. With the pod's supply of rocket fuel exhausted, Ivan has no choice but to drop directly down into the black hole.

What does Veronica see? She sees Ivan and the pod fall together, moving faster and faster. She sees Ivan's heart beat more and more slowly—both because he is lower than her (gravitational time dilation) and because he is moving (special relativistic time dilation). She sees the pod turn from green to orange to red to infrared.

But then, as she watches Ivan's pod through her infrared detecting goggles, she notices that it's no longer moving faster and faster. Because everything down there proceeds slowly, the pod moves slowly. She sees Ivan's pod slow almost to a halt before it reaches the event horizon. The oscillations from Ivan's atoms become so slow that even her infrared detector can't detect them. Ivan is lost to her, but for the rest of her life she will remember her last infrared view of Ivan's pod, ever so slowly creeping toward the event horizon.

What does Denise see? As she hovers just above the event horizon, she sees Ivan's pod plummet down, going faster and faster. Her heart is already beating slowly, relative to Veronica's, so when Ivan's pod passes by her she notices only the special relativistic time dilation, not the gravitational time dilation. Denise does not see the pod slow to a crawl before reaching the event horizon. Instead, she sees it plunge into and through the event horizon moving at nearly the speed of light.

Finally, what does Ivan see and feel? After the escape pod's rockets complete their final blast, Ivan and the pod fall toward the black hole at the same rate, so Ivan is "weightless" in the sense that we've used before. As he falls, he sees for himself the same phenomena that Denise described: the growing bulk of the inky blackness, the distortion of the constellations, the view of his own pod as its light orbits the black hole, the mother ship orbiting faster and faster, the mother ship turning from green to blue to violet to ultraviolet. And then, he notices himself fall past Denise's shuttle and through the event horizon. He feels no jolt, no shudder, as he passes through the event horizon: he is, as always, weightless because (principle of equivalence) the pod and his body accelerate downward at the same rate.

Ivan looks around: the inky blackness has gone away. Light from stars all around falls into the black hole, and he can see that light using his ultraviolet detecting goggles. The light follows intricately curved paths, so the constellations are still distorted. Ivan knows that he can never send a message out to Denise or Veronica, but he can still watch them, through his ultraviolet goggles. All of their actions are to him very fast, and in fact he sees not only their entire lives, but he watches with interest the lives of their children and grandchildren.

All proceeds well until Ivan, dropping feet-first toward the center of the black hole, feels a small tug on his feet. He's been weightless for so long that he hardly recognizes the feeling. At issue is not the intensity of the gravity: If Denise had entered the event horizon, rockets blasting to keep herself almost stationary with respect to the hole, then she would have felt great weight from the very intense gravity under the event horizon. But Denise declined to enter. Ivan has taken the different tack of falling through the event horizon and thus remaining weightless. The intensity of the gravity affects him not at all. But what *does* affect Ivan is the *change* in the intensity of gravity over the length of his body: the gravity at his feet is stronger than the gravity at his head. Both his feet and his head fall, but his feet fall at a greater rate. The change of intensity increases as Ivan falls. At first his bones hold his feet and head together, but eventually nothing can stop the tugging: Ivan's body tears into shreds.

Do not mourn for Ivan. We must all die, and Ivan selected a black hole where the shredding process takes place so rapidly that it was finished before the nerve signals even reached his brain, so his death was painless. And during his fall, Ivan managed to follow the lives of several generations on the outside, and to experience for himself the strange and beautiful mysteries of our home, the universe.

REFERENCES

The assertions made in this chapter are backed up by explanations and detailed treatments in books like

Edwin F. Taylor and John Archibald Wheeler, *Exploring Black Holes: Introduction to General Relativity* (San Francisco: Addison Wesley Longman, 2000).
James B. Hartle, *Gravity: An Introduction to Einstein's General Relativity* (San Francisco: Addison-Wesley, 2003).
Kip S. Thorne, *Black Holes and Time Warps: Einstein's Outrageous Legacy* (New York: W. W. Norton, 1994).
Robert Geroch, *General Relativity from A to B* (Chicago: University of Chicago Press, 1981).

Splendid visualizations of the phenomena described in this chapter are available through these web sites:

Andrew Hamilton, "Falling Into a Black Hole" http://casa.colorado.edu/~ajsh/schw.shtml.
Ute Kraus and Corvin Zahn, "Relativity Visualized: Space Time Travel" www.spacetimetravel.org/.
Robert Nemiroff, "Virtual Trips to Black Holes and Neutron Stars" http://antwrp.gsfc.nasa.gov/htmltest/rjn_bht.html.

CHAPTER TWENTY

The Vista Open to Us

W hat a long, strange journey it has been!

It started out so innocently, asking questions about mirrors in railroad coaches. But those innocent questions led, through thought, doubt, and the harsh refining light of experiment, to a whole new understanding of space and time. It's easy to wrap that new knowledge up into a few phrases—"a moving clock ticks slowly," "a moving rod is short," "the rear clock is set ahead"—just as it's easy to learn a few stock phrases in a foreign language. But trying to grow familiar with this new knowledge results in surprising and unexpected consequences—"no body is completely rigid," "a rod can be straight in one reference frame and bent in another"—just as there are surprises and unexpected consequences in growing familiar with a foreign culture.

Please don't think that just because you've reached the end of the book, you've reached the end of the surprises. I've been studying relativity since 1973, and it still surprises me. In this last chapter I'll set out some of the paths that you could take from here, if you want more surprises.

Leonardo da Vinci maintained that "all those sciences are vain and full of error that are not born of experience, mother of all certainty." Experimental relativity is full of challenges and full of insights. Here are two challenges: (1) We've seen that in 1972 time dilation was tested directly using two atomic clocks, one moving relative to the other. Atomic clocks have improved enormously since 1972, and this experiment should be repeated. (2) We have no direct experimental test of length contraction. Can you think of one? Scientists are hard at work on these and related issues: the easiest way to track their progress is through the Inside Science News Service at InsideScience.org.

In this book we've used a minimum of mathematics so that we don't confuse working the formalism with understanding nature. Yet the mathematical formalism is itself both useful and beautiful. The Lorentz transformation, four-vectors, tensors, the Poincaré group—the names alone make me shiver in delight. When properly used, these mathematical tools can help you solve problems *and* help you appreciate the universe. Among the many first-rate textbooks that could launch you down an exploration of this path, I particularly recommend

Thomas A. Moore, *Six Ideas That Shaped Physics, Unit R: The Laws of Physics are Frame-Independent*, 2nd ed. (New York: McGraw-Hill, 2003).

If relativity changes our ideas of space and time, will it change our ideas about how bodies move in space and time? Of course! We need new ideas about energy, momentum, force, mass. One surprise uncovered when walking down this path is that energy and mass are related through $E = mc^2$, but that's not the only surprise. To anyone wishing to explore this path, I again recommend Moore's textbook.

The principle of relativity, plus the rule that no causal signal travels faster than light, put great constraints on the kinds of forces that one body can exert on another. There are, for example, constraints on the behavior of electric forces, constraints on the behavior of magnetic forces, and so forth. The disciple of relativistic field theory looks into these constraints and investigates the character of forces that actually occur in nature. This path is deep, useful, and intrinsically mathematical. It requires considerable background study to understand even the first page of

Davison E. Soper, *Classical Field Theory* (Wiley, New York, 1976),

but simply running your eye over the pages will give you a feel for the mathematical elegance of this path.

Chapters 17 through 19 this book introduced general relativity—the relativistic theory of gravity. One surprise down this path is the black hole. Others include gravitational radiation and the expansion of the universe. Still other surprises, I am convinced, lie covered, awaiting discovery. Sketchy introductions to this broad set of intertwined paths are provided by

Marcia Bartusiak, *Einstein's Unfinished Symphony: Listening to the Sounds of Space-Time* (Washington, DC: Joseph Henry Press, 2000).

Robert M. Wald, *Space, Time, and Gravity: The Theory of the Big Bang and Black Holes,* 2nd ed. (Chicago: University of Chicago Press, 1992).

Many of the tenets of general relativity find their best application, and their most stringent tests, in the world of astronomy. I recommend the Astronomy Picture of the Day Web site as a beautiful and rigorous way to keep up with developments in this field.

Relativistic thinking has opened new paths in our understanding of nature and also in our understanding of politics, anthropology, the arts, culture, and society in general. Sometimes it seems that misconceptions about relativity have had more influence than relativity itself has. Such influences are discussed in

L. Pearce Williams, *Relativity Theory: Its Origins and Impact on Modern Thought* (New York: Wiley, 1968).

Harry Woolf, ed., *Some Strangeness in the Proportion: A Centennial Symposium to Celebrate the Achievements of Albert Einstein* (Reading, MA: Addison-Wesley, 1980).

Gerald Holton and Yehuda Elkana, eds., *Albert Einstein: Historical and Cultural Perspectives* (Princeton: Princeton University Press, 1982).

Iain Paul, *Science, Theology, and Einstein* (New York: Oxford University Press, 1982).

Dennis P. Ryan, *Einstein and the Humanities* (Westport, CT: Greenwood, 1987).

They are also demonstrated in the novel

Alan Lightman, *Einstein's Dreams* (New York: Pantheon Books, 1993).

Or perhaps you're interested, not in where we can take our ideas about relativity, but in where those ideas came from. The history of relativistic thinking is full of extraordinary insights, simple blunders, and blind alleys. Because Albert Einstein played such a large role in this history, it's easy to think that he did everything by himself. You'll realize that he didn't after reading

Banesh Hoffmann, *Relativity and Its Roots* (New York: Scientific American Books, 1983).

Loyd S. Swenson, *Genesis of Relativity: Einstein in Context* (New York: B. Franklin, 1979).

Nevertheless, Einstein's importance demands that I list his definitive, scholarly biography,

Albrecht Fölsing, *Albert Einstein: A Biography* (New York: Viking, 1997),

as well as the more popular

Walter Isaacson, *Einstein: His Life and Universe* (New York: Simon and Schuster, 2007).

Isaacson is neither a physicist nor a historian, but a journalist.* He does a particularly good job in establishing the truth about certain rumors concerning Einstein:
both rumors that he was saintly and rumors that he was the opposite. Despite the
coquettish title,

Dennis Overbye, *Einstein in Love: A Scientific Romance* (New York: Viking, 2000),

is a reliable biography of both the personal and scientific aspects of Einstein's first
40 years. Finally, no one should miss

Jürgen Renn and R. Schulmann, eds., *Albert Einstein/Mileva Maric: The Love Letters*
    (Princeton, NJ: Princeton University Press, 1992),

which makes a great Valentine's Day gift.

Projects

20.1 Project: How do you feel? Science is not affected by our personal reaction to nature,
so students of science are usually discouraged from discussing these reactions. Yet it is
inevitable that these reactions occur, whether we discuss them or not. It is better to
acknowledge them than to pretend that they don't exist. Write an essay describing your
personal reaction to the science that you learned through reading this book. Does it make
you feel morose? Sorrowful? Or merely confused? Do you find it elating? Magnificent?
Mind-expanding or mind-destroying?

For comparison, you might find it interesting that one of the greatest scientists of the
nineteenth century, Michael Faraday, wrote in his laboratory diary (entry 10040, 19 March
1849) that "nothing is too wonderful to be true."

20.2 Project: Relativity and the law. The U.S. Constitution prohibits *ex post facto* laws. That
is, you can't be arrested for an act that you performed before the law prohibiting that
action was enacted. The sequence of events must be "enactment first, illegal act second."
Because of the relativity of simultaneity the sequence of events might be different from
one reference frame to another. The Constitution doesn't specify reference frame, so
from a relativistic perspective this provision is dangerously vague.

Read through your favorite law (perhaps your country's constitution) in detail. Identify
and recraft each clause that would require change if travel close to light speed became
commonplace.

---

*On pp. 148 and 349 he errs in describing gravitational time dilation: he claims that clocks in
more intense gravitational field tick slowly, whereas in fact it is lower clocks tick that slowly, even
when the field is uniform.

20.3 Project: Relativity and the language. I mentioned on page 45 that the English language was not built to correctly echo relativistic concepts of space and time. Design a relativistically appropriate language. The most obvious problem to tackle is verb tense, but there are others.

Appendix A

For the Cognoscenti

Why did I make the choices I made in writing this book?

Where Is the Lorentz Transformation?

I begin with an analogy: A baby begins to learn about rotation when he first rolls over in his crib. Later—much later—that baby might learn about orthogonal transformations, rotation matrices, quaternions, Euler angles, irreducible representations of the rotation group, the Wigner-Eckart Theorem, and so forth. I don't need to tell you that these are very powerful ideas that solve complicated problems with ease. And I don't need to tell you that they're also very beautiful ideas. But their beauty appeals to the intellect and not to the gut. They're not the best way to give one a qualitative sense of what's going on in rotation: you're lying on your side and want to roll over onto your tummy, if you don't consider the rotation matrix.

As with rotations, so with relativity. We all know and love the Lorentz transformations. We know how powerful they are, how beautiful they are. But we also know that they're hard to understand at a gut level.

What, if anything, is the "natural, gut-feeling" analog of the Lorentz transformation? I maintain that they are the three principles emphasized in this book: time dilation, length contraction, and the relativity of simultaneity. These principles are the best way to approach problems for qualitative insight, whether that insight is the final goal or a preliminary to a more formal computation.

I readily concede that using the Lorentz transformation is less prone to error, more automatic, and smoother than using the three principles—and this is precisely why the three principles are valuable. The Lorentz transformations are so smooth and powerful that they can be used blindly without understanding what's going on.

A disadvantage of this approach is that students can get the misimpression that relativity is about devious rods and malfunctioning clocks rather than about space and time. To counteract this tendency, I start off talking about clocks and slowly slide into talking about time. (For example, I start off talking about "the relativity of synchronization" and end up talking about "the relativity of simultaneity.")

Conventions and Word Choice

You should be aware of certain conventions to which I adhere, and the rationale behind those conventions:

A moving clock "ticks slowly" rather than "runs slowly," to avoid confusion between the clock running and the observer moving because she is running.

Clocks are located "front" and "rear," and their time readings are "ahead" and "behind." I never say that one clock is located "behind" another.

I avoid the words "appears," "views," and "measures," in favor of "is," "observes," and "sees" (see page 69). To "observe" means to figure out what's really going on, either by accounting for light-travel delays, or by using a "latticework of clocks" [E.F. Taylor and J.A. Wheeler, *Spacetime Physics* (Freeman, New York, 1992)], or by using the "parallel racetrack" (page 69). To "see" means to look with your eye using the old light striking your eye. While an observation depends upon reference frame, a "seeing" depends upon both reference frame and the location of the looker.

Situations usually involve one male and one female character, so that pronouns are unambiguous.

Light Propagation Time

Most books at this level don't devote attention to light propagation time and the consequent difference between what is seen and what is observed. The general opinion seems to be that this is an advanced topic and that if authors don't mention the difference, readers won't think about it and hence won't get confused. In my experience this attitude is dead wrong. Readers think about it whether or not authors bring it up, and when they do, they usually think about it incorrectly. Many times I have heard advanced physics students confidently explain to neophytes that "moving clocks don't really tick slowly, but the old light reaching your eye makes your image of the clock appear to tick slowly." [My experience has been borne out by physics education research: see Rachel E. Scherr, Peter S. Shaffer, and Stamatis Vokos, "The challenge of changing deeply held student beliefs about the relativity of simultaneity," *American Journal of Physics*, 70 (2002) 1238–48.] Rather than allow such misconceptions to fester, I attack them head on.

Ranks, Numbers, and Symbols

I think of statements, arguments, examples, and problems as falling into three categories: ranking, numerical, and symbolic. A ranking statement is "a moving rod is short." A numerical statement is "a rod of rest length 15 feet, moving at speed $\frac{3}{5}c$, has length 12 feet." A symbolic statement is "a rod of rest length L_0, moving at speed V, has length $\sqrt{1 - (V/c)^2}L_0$."

Each kind of statement has its own pedagogical role to play. This book is intended for a wide range of readers, and certainly some readers will be more comfortable with one approach than with others. But *every* reader will attain deeper understanding when all three approaches are used. This book does not take one approach; instead, different topics are approached using whichever of the three attack lines I have found to be most effective for that topic. Important topics are approached using all three lines.

This book often uses specific numbers in places where a book for physicists alone would use algebraic symbols. Why? Physicists like us value generality and abstractness for their power. Most readers of this book, in contrast, are suspicious of generality and abstractness. And they should be! They know all too well that generality is often an excuse for vagueness, a mask for ignorance, an invitation for absurd interpretations, and a curtain to hide unpleasant specifics. Anyone who has ever purchased a used car (or voted for a politician) is apt to prefer firm specifics over vague generalities.

Americanisms

I teach in the United States of American to a largely American audience, so I use analogies and units that Americans find familiar. For example, I talk about using a flat map to navigate from Detroit to Cleveland, and about a pole that's 30 feet long. If I were presenting this material in Australia, I would instead talk about using a flat map to navigate from Sidney to Canberra, and about a pole that's 10 meters long.

My audience finds the use of SI units to be stuffy and pretentious. An Australian audience would find the use of feet to be stuffy and pretentious. In order to aid conversion of this book's arguments in countries where SI units are commonplace, every example and problem (except in chapter 9, "He Says, She Says," the section on measuring the length of a moving rod, and in chapter 14, "The Pole in the Barn") uses lengths that are integral multiples of 3 feet, so the 60-foot railroad coach can be readily converted to the 20-meter railroad coach. I recommend that teachers in SI countries use as a unit of time the "nanne," such that 1 nanne is the time required for light to travel 1/3 meter.* Thus, the speed of light is exactly 1/3 meter/nanne, and 1 nanne is about 1.11 nanoseconds.

What Is Science?

Finally, to my surprise I've found that even students who have a good grasp of science often have a poor understanding of the character and the limitations of science.[†] Surely, it is more important to know the difference between fact and opinion than it is to know that moving clocks tick slowly. For this reason, I've treated the character of science not only implicitly (by treating scientific topics in a scientific manner) but also explicitly.

Some disciplines attempt to find final truths, applicable for all time and to all situations. You can judge for yourself whether they succeed. Science tries to find out what it can today, acknowledging our ignorance and limitations and errors. Tomorrow we'll know more than we do today, so the generalities that work today might not work

---

*Word origin: *nanne* is to *nan* as *tonne* is to *ton*.

[†]Their understanding isn't helped by policy advocates who use the term "sound science" to mean "any evidence, however flimsy, that supports my predetermined position." Nor does it help that creationists consider any tentative result (and hence any scientific result) to be "controversial."

tomorrow. As Galileo put it in 1632, "Geometrical exactitude should not be sought in physical matters."*

Yet while science is tentative and limited, it is also reliable. Once again Einstein put it better than I can: "All our science, measured against reality, is primitive and childlike— and yet it is the most precious thing we have."†

Dialogue Concerning the Two Chief World Systems, translated by Stillman Drake (The Modern Library, New York, 2001) page 15, postil.

†Letter to Hans Muehsam, 9 July 1951. Alice Calaprice, *The New Quotable Einstein* (Princeton University Press, 2005) page 245.

Appendix B

Hints

Before using these hints, be sure to read page 9 about the principles behind active learning and problem solving.

1.2. **Distance traveled in various frames.** Use distance traveled = speed × time elapsed, but realize that the speed is different in different frames.

2.1. **Backseat shooting.** (b) As in problem 1.2, the distance traveled depends on reference frame.

3.1. **Looking at a clock.** It takes some time for light to travel from the clock to Cynthia's eye. How much time?

4.2. **Time dilation examples.** In these examples we are given $T = 1$ hr, the time elapsed in our frame, and asked to find T_0, the time ticked off by a moving clock.

4.4. **Muon lifetime.** In the laboratory frame, the lifetime of such a muon is 3.9 microseconds. (That moving muon decays slowly.)

5.1. **Another race.** This is the same situation analyzed in chapter 5, "The Great Race," but with different numbers.

7.1. **The meaning of "rear."** Reread the section "What Does 'Rear' Mean?" on page 52. You will find it helpful to sketch the situation in the Earth's frame, in the sports car's frame, and in the sedan's frame.

7.2. **Engine trouble.** Suppose there were a clock by the driver, and another clock by the tail pipe, and that these two clocks were synchronized in the car's frame.

7.5. **Boxcar with a bomb.** The two flashlights flash simultaneously in the boxcar's frame. Is this true in the Earth's frame?

7.6. **Train in a tunnel.** The resolution requires one sketch in the train's frame and two in the tunnel's frame.

8.1. **A trip to the grocery store.** (a) No relativity needed to solve this part, just the definition speed = distance traveled/time elapsed. (b) Time dilation: Tabetha's moving wristwatch ticks slowly. (c) Length contraction: a moving rod (or a moving slab of highway pavement) is short. (e) Time dilation again. (f) Relativity of synchronization.

8.2. **Two fast women.** In Veronica's frame the right-hand (rear) clock is set ahead by 9 nan. In Denise's frame the right-hand (rear) clock is set ahead by 12 nan. And in both frames, these moving clocks tick slowly.

8.3.   **To see and to observe.** Recall the difference between "see" and "observe" described on page 69.

8.4.   **Racetrack.** (c) Use symmetry between the racetrack's frame and Alia's frame. The two frames differ only in that Alia moves to the right in the racetrack's frame, whereas the racetrack moves to the left in Alia's frame. (d) Solve this problem using speed = distance traveled/time elapsed. No relativity is required.

9.2.   **Veronica peers ahead.** Compare this situation with that of Aaron watching Rosemary's home clock with binoculars (see page 68).

9.4.   **A chorus line.** In the car's reference frame, the girls are moving left, so the rear girl is the one on the right. (If this isn't clear, sketch the situation from the car's reference frame.)

9.6.   **Heartbeat.** Veronica's heart still beats at one beat per second *in Veronica's own frame.*

10.1.   **Faster than light.** Is the motion of the red dot a causal signal?

14.1.   **Cloudburst.** In the barn's frame, rain falls simultaneously to the left and right of the barn. Are those two rainfalls simultaneous in the vaulter's frame?

14.2.   **Fatal mistake.** How rigid is that pole?

14.3.   **Electrical connection.** The electrical signal travels at the speed of light. In the barn's frame, when does the back door open?

14.4.   **Chain connection.** Is the chain perfectly stiff, or is it stretchy and flexible like a rubber band?

15.2.   Use the formula Veronica derived on page 133, just before her face glowed with excitement, but use "8 yr" instead of "200 yr."

16.1.   **Relativistic baking.** In the baker's frame, the front and rear edges of the cookie cutter hit the dough simultaneously. In the dough's frame, . . .

16.8.   **Busted bus.** In the Earth's frame, the front and rear ends of the bus are crushed simultaneously. In the bus's frame, . . .

16.9.   **Falling hula hoop.** In the Earth's frame, the hula hoop is horizontal. In the yardstick's frame the hula hoop is tilted. To answer the question you'll need to explain why and in which direction it's tilted.

16.10.   **Explosion!** How hard is the "hardened steel"?

16.11.   **Stealth tank.** Is the tank rigid? Is it straight in all reference frames?

16.13.   **Joust.** Are the lances rigid? (Compare problem 16.12.)

Appendix C

Skeleton Answers

Before using these skeleton answers, be sure to read page 9 about the principles behind active learning and problem solving. And before you copy these skeleton answers into your homework assignment, recall the difference between a solution and a skeleton answer.

1.1. **Speed in various frames.** (a) 45 miles/hour. (b) 0 mile/hour. *Moral:* The speed is not an "absolute"—the same in all reference frames. The speed is different in different frames.

1.2. **Distance traveled in various frames.** (a) 40 miles. (b) 30 miles. (c) 0 mile. *Moral:* The distance traveled by a car is not an "absolute"—the same in all reference frames. The distance traveled is different in different frames.

2.1. **Backseat shooting.** (a) Northeast; between north and northeast. *Moral:* The direction traveled by a dart is not an "absolute"—the same in all reference frames. The direction traveled is different in different frames.

3.1. **Looking at a clock.** 473 nan. *Moral:* What you see is what happened some time ago because it takes time for light to reach your eye.

4.1. **Finding the square root factor.** (b) 0.9942. *Moral:* The speed 20 000 miles per second is very high by everyday standards, yet so much smaller than speed of light that the time dilation factor is nearly equal to 1.

4.3. **Time dilation for a biological clock.** 10 weeks. *Moral:* All moving clocks tick slowly: light clocks, pendulum clocks, biological clocks, etc.

4.4. **Muon lifetime.** (a) 550 meters. (b) 980 meters.

5.1. **Another race.** (a) 507 nan. (b) 468 nan. (c) 180 ft. (d) 432 nan. (e) 075 nan. *Moral:* The phenomena of length contraction and relativity of synchronization occur for a race of any speed and a racetrack of any length, not just for the special values used in chapter 5, "The Great Race."

7.1. **The meaning of "rear."** (a) Trunk is to the rear. (b) Hood is to the rear. (c) Neither trunk nor hood is to the rear.

7.2. **Engine trouble.** (b) Because the rear event happens first.

7.3. **Three clocks.** B: 106 nan. C: 118 nan. *Moral:* You get the same answer for clock C whether you step from A to B to C, or go directly from A to C.

7.4. **Two clocks.** 94 nan. *Moral:* "Rear clock set ahead" means the same as "front clock set behind."

7.5. **Boxcar with a bomb.** In the Earth's frame, the rear light signal does indeed have farther to travel than the front light signal. But the rear clock is set ahead, so the rear light signal starts out earlier. The bomb does not explode. *Moral:* It is not a paradox that the two clocks aren't synchronized in the Earth's frame; indeed, it *would* be a paradox if they *were* synchronized!

7.7. **Veronica in a semitrailer.** (a) 62 nan. (b) 86 nan. (c) 48 ft.

8.1. **A trip to the grocery store.** (a) 1500 nan. (b) 900 nan. (c) 720 ft. (d) 900 nan, same as in part (b). (Relativity really does make sense; it just doesn't make common sense.) (e) 540 nan. (f) In Tabetha's frame, the store clock is set ahead of the home clock by $L_0 V/c^2 = 960$ nan. When she starts her journey, the store clock reads (in her frame) 960 nan rather than 000 nan. A clock that starts off reading 960 nan and then ticks off 540 nan of course ends up reading 1500 nan.

8.2. **Two fast women.** 0 nan and –5 nan. *Moral:* Two events can occur in one sequence in one reference frame, simultaneously in another, and in the opposite sequence in a third. The sequence of these two events is *not* an "absolute"—i.e., not the same in all reference frames.

8.3. **To see and to observe.** At 170 nans after noon.

8.4. **Racetrack.** (a) 54 ft. (b) 72 nan. (d) 112.5 nan. (e) 67.5 nan.

9.2. **Veronica peers ahead.** 20 nan.

9.3. **American graffiti.** (a) Jennifer. (b) 90 nan. (c) 54 ft. (d) 150 ft.

9.4. **A chorus line.** The girl on the far right kicks first, then the girl left of her, then the girl left of *her*, and so forth until the girl on the far left kicks. A photographer at rest in the car's frame would need a video camera.

9.6. **Heartbeat.** Veronica's cells are respiring slowly, also.

9.7. **Racetrack.** (a) 24 ft. (b) A: 18 nan. (c) B: 40 nan, C: 32 nan, D: 50 nan. (d) 50 nan. (e) 50 nan. (f) 40 nan. *Moral:* The moral of this problem is the same as the moral of chapter 9, "He Says, She Says." Ivan and Veronica disagree on whether clocks are synchronized and on which clock ticks slowly. But they agree on the reading of Ivan's clock and the reading of Veronica's clock as the two clocks pass each other. Everything's okay: The two differ not in what the passing clocks read but only in their explanation for why those clocks read as they do. They can maintain peace by agreeing to disagree. In short, relativity makes sense. It doesn't make common sense, but it makes sense!

10.1. **Faster than light.** In the last experiment, the red dot doesn't begin to move until 1 second after I flick my wrist.

11.1. **Speed addition.** (a) $\frac{4}{5}c$. (b) $\frac{16}{19}c$.

11.2. **Two fast women.** $+\frac{5}{13}c$; $-\frac{5}{13}c$.

11.3. **High-speed approach.** If common sense were correct, Carla would approach Edwin at $-1\frac{1}{4}c$, faster than light speed. But in fact she approaches him at $-\frac{10}{11}c$, less than light speed.

11.4. **Tennis in a train.** In the Earth's frame everything happening within the railroad coach happens slowly (time dilation). Furthermore, the coach itself is short (length contraction), so when the ball moves from one end of the coach to the other, it doesn't move very far. If it moves a short distance in a long time, it moves slowly!

11.6. **Which clock set which way?** It's the *rear* clock that's set ahead.

12.1. **Faster than light.** I assumed that the stick was rigid.

13.2. **The return of the hungry traveler.** During Rosemary's turnaround, the home clock jumps ahead by 108 nans.

13.3. **Time travel.** $V = \frac{\sqrt{35}}{6} c \approx 0.986\, c$.

13.5. **Speedup.** Before the speedup, Earth's calendar is set ahead of Sirius's calendar by $L_0 V/c^2 = (8.0\ \text{yr})c(\frac{4}{5}c)/c^2 = 6.4$ yr, so it reads 3.6 years. After the speedup, Earth's calendar is set ahead of Sirius's calendar by $L_0 V/c^2 = (8.0\ \text{yr})c(\frac{9}{10}c)/c^2 = 7.2$ yr, so it reads 2.8 years. Earth's calendar jumps backwards! Nevertheless, if a baby is born on Earth 3.0 years after Veronica's departure, she will *not* see its birth twice because it takes time for the light from the baby's birth to reach Veronica's eye.

14.1. **Cloudburst.** Dry.

14.2. **Fatal mistake.** Closed.

14.3. **Electrical connection.** The back door would open when its clock reads 175 nan. But the tip of the pole already smashed through this closed door back when that clock read 125 nan.

15.3. $V = \frac{1}{\sqrt{2}} c$.

15.5. **Travel at the speed of light.** As Phil Photon was absorbed within the star Spica, he thought back over his life: "Spica traveled to me over zero length, and it required zero time!" (Duh.)

Appendix D

Ready Reference

Values for the Square Root Factor

V	$\sqrt{1-(V/c)^2}$	V	$\sqrt{1-(V/c)^2}$
$\frac{1}{5}c$	$\frac{\sqrt{24}}{5} \approx 0.98$	$\frac{8}{17}c$	$\frac{15}{17}$
$\frac{2}{5}c$	$\frac{\sqrt{21}}{5} \approx 0.92$	$\frac{15}{17}c$	$\frac{8}{17}$
$\frac{3}{5}c$	$\frac{4}{5} = 0.80$	$\frac{12}{37}c$	$\frac{35}{37}$
$\frac{4}{5}c$	$\frac{3}{5} = 0.60$	$\frac{35}{37}c$	$\frac{12}{37}$
$\frac{5}{13}c$	$\frac{12}{13}$	$0.28\,c$	0.96
$\frac{12}{13}c$	$\frac{5}{13}$	$0.96\,c$	0.28

Summary of Special Relativity

Time dilation	A moving clock ticks slowly.	$T = \dfrac{T_0}{\sqrt{1-(V/c)^2}}$

T_0 is the time ticked off by a single moving clock (which is also the time elapsed in that clock's own frame).
T is the (longer) time elapsed in the frame in which that clock moves at speed V.

Length contraction	A moving rod is short.	$L = L_0\sqrt{1-(V/c)^2}$

L_0 is the length of a rod in that rod's own frame (its "rest length").
L is the (shorter) length of that rod in the frame in which that rod moves at speed V.

Relativity of synchronization	A moving pair of clocks isn't synchronized.	Rear clock set ahead by L_0V/c^2.
Also called **relativity of simultaneity**	If two events are simultaneous in one frame, then in another frame the rear event happens first.	

If a pair of clocks is synchronized in that pair's own frame, then in the frame in which they both move at speed V, the rear (trailing) clock is set ahead by L_0V/c^2.

Problem-Solving Tips

1. Keep your reference frame in mind. A clock moving in one frame is stationary in another. A pair of clocks stationary in one frame (and synchronized) is moving in another (and hence not synchronized). If you're not sure which frame you're using, then you won't be sure of which clocks tick slowly, which rods are short, and which clocks are synchronized.

2. T **versus** T_0. In the clock's own frame, the time ticked off is the same as the time elapsed. That's why a clock is useful. But if the clock is moving, it ticks slowly: the time elapsed (T) is more than the time ticked off (T_0). The clock is still useful, but you have to use the formula to get from the time ticked off (what you measure) to the time elapsed (what you want to know).

3. Sketch. Draw at least two sketches for each problem, one from each reference frame. But because two events that are simultaneous in one frame might not be simultaneous in another, you might need to draw one sketch from one frame and two sketches from another frame.

4. What you see is not what is happening. It takes some time for light to move from one place to another.

5. Lack of rigidity. All objects are compressible and flexible like playdough.

Index